REGISTRY
OF
MAINE
TOOLMAKERS

Hand Tools in History Series

- Volume 6: Steel- and Toolmaking Strategies and Techniques before 1870
- Volume 7: Art of the Edge Tool: The Ferrous Metallurgy of New England Shipsmiths and Toolmakers from the Construction of Maine's First Ship, the Pinnace *Virginia* (1607), to 1882
- Volume 8: The Classic Period of American Toolmaking, 1827-1930
- Volume 9: Davistown Museum Exhibition: An Archaeology of Tools
- Volume 10: Registry of Maine Toolmakers
- Volume 11: Handbook for Ironmongers: A Glossary of Ferrous Metallurgy Terms: A Voyage through the Labyrinth of Steel- and Toolmaking Strategies and Techniques 2000 BC to 1950

REGISTRY
OF
MAINE
TOOLMAKERS

A Compilation of Toolmakers Working in Maine and the Province of Maine Prior to 1900

Including information on the
Robert Merchant Wantage Rule, Berwick, Maine, 1720,
the oldest signed and dated measuring tool made in Maine
and
Early Maine Planemakers
Joseph Metcalf and Thomas Waterman

H. G. Brack

Davistown Museum
Publication Series Volume 10

© Davistown Museum 2008
ISBN 978-0-9769153-0-0

ISBN 9: 0-9769153-0-8
ISBN 13: 978-0-9769153-0-0
Davistown Museum

5th edition

Front cover illustration:
Robert Merchant wantage rule from The Davistown Museum MI collection (ID# TBW1006).

Cover design by Sett Balise

This publication was made possible by a donation from Barker Steel LLC.

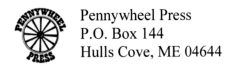

Pennywheel Press
P.O. Box 144
Hulls Cove, ME 04644

Preface

Davistown Museum *Hand Tools in History*

One of the primary missions of the Davistown Museum is the recovery, preservation, interpretation, and display of the hand tools of the maritime culture of Maine and New England (1607-1900). The *Hand Tools in History* series, sponsored by the museum's Center for the Study of Early Tools, plays a vital role in achieving the museum mission by documenting and interpreting the history, science, and art of toolmaking. The Davistown Museum combines the *Hand Tools in History* publication series, its exhibition of hand tools, and bibliographic, library, and website resources to construct an historical overview of steelmaking techniques and strategies and the edge toolmakers of New England's wooden age. Included in this overview are the roots of these strategies and techniques in the early Iron Age, their relationship with modern steelmaking technologies, and their culmination in the florescence of American hand tool manufacturing in the last half of the 19th century.

Background

During 38 years of searching for New England's old woodworking tools for his Jonesport Wood Company stores, curator and series author H. G. Skip Brack collected many different tool forms with numerous variations in metallurgical composition, many signed by their makers. The recurrent discovery of forge welded tools made in the 18th and 19th centuries provided the impetus for founding the museum and then researching and writing the *Hand Tools in History* publications. In studying the tools in the museum collection, Brack found that, in many cases, the tools seemed to contradict the popularly held belief that all shipwrights' tools and other edge tools used before the Civil War originated from Sheffield and other English tool-producing centers. In many cases, the tools that he recovered from New England tool chests and collections dating from before 1860 appeared to be American-made rather than imported from English tool-producing centers. Brack's observations and the questions that arose from them led him to research the topic and then to share his findings in the *Hand Tools in History* series.

Hand Tools in History Publications

- Volume 6: *Steel- and Toolmaking Strategies and Techniques before 1870* explores ancient and early modern steel- and toolmaking strategies and techniques, including those of early Iron Age, Roman, medieval, and Renaissance metallurgists and toolmakers. Also reviewed are the technological innovations of the Industrial Revolution, the contributions of the English industrial revolutionaries to the evolution of the factory system of mass production with interchangeable parts, and the development of bulk steelmaking processes and alloy steel technologies in the latter half of the 19th century. Many of these technologies play a role in the florescence of American ironmongers and toolmakers in the 18th and 19th century. Author H. G. Skip Brack cites archaeometallurgists such as Barraclough, Tylecote, Tweedle, Smith, Wertime, Wayman, and many others as useful guides for a journey through the pyrotechnics of ancient and modern metallurgy. Volume 6 includes an extensive bibliography of resources pertaining to steel- and toolmaking techniques from the early Bronze Age to the beginning of bulk-processed steel production after 1870.

- Volume 7: *Art of the Edge Tool: The Ferrous Metallurgy of New England Shipsmiths and Toolmakers* explores the evolution of tool- and steelmaking techniques by New England's shipsmiths and edge toolmakers from 1607-1882. This volume uses the construction of Maine's first ship, the pinnace *Virginia*, at Fort St. George on the Kennebec River in Maine (1607-1608), as the iconic beginning of a critically important component of colonial and early American history. While there were hundreds of small shallops and pinnaces built in North and South America by French, English, Spanish, and other explorers before 1607, the construction of the *Virginia* symbolizes the very beginning of New England's three centuries of wooden shipbuilding. This volume explores the links between the construction of the *Virginia* and the later flowering of the colonial iron industry; the relationship of 17th, 18th, and 19th century edge toolmaking techniques to the steelmaking strategies of the Renaissance; and the roots of America's indigenous iron industry in the bog iron deposits of southeastern Massachusetts and the many forges and furnaces that were built there in the early colonial period. It explores and explains this milieu, which forms the context for the productivity of New England's many shipsmiths and edge toolmakers, including the final flowering of shipbuilding in Maine in the 19th century. Also included is a bibliography of sources cited in the text.

- Volume 8: *The Classic Period of American Toolmaking 1827-1930* considers the wide variety of toolmaking industries that arose after the colonial period and its robust tradition of edge toolmaking. It discusses the origins of the florescence of American toolmaking not only in English and continental traditions, which produced gorgeous hand tools in the 18th and 19th centuries, but also in the poorly documented and often unacknowledged work of New England shipsmiths, blacksmiths, and toolmakers. This volume explicates the success of the innovative American factory system, illustrated by an ever-expanding repertoire of iron- and steelmaking strategies and the widening variety of tools produced by this factory system. It traces the vigorous growth of an American hand toolmaking industry that was based on a rapidly expanding economy, the rich natural resources of North America, and continuous westward expansion until the late 19th century. It also includes a company by company synopsis of America's most important hand toolmakers working before 1900, an extensive bibliography of sources that deal with the Industrial Revolution in America, special topic bibliographies on a variety of trades, and a timeline of the most important developments in this toolmaking florescence.

- Volume 9: *An Archaeology of Tools* contains the ever-expanding list of tools in the Davistown Museum collection, which includes important tools from many sources. The tools in the museum exhibition and school loan program that are listed in Volume 9 serve as a primary resource for information about the diversity of tool- and steelmaking strategies and techniques and the locations of manufacturers of the tools used by American artisans from the colonial period until the late 19th century.

- Volume 10: *Registry of Maine Toolmakers* fulfills an important part of the mission of the Center for the Study of Early Tools, i.e. the documentation of the Maine toolmakers and planemakers working in Maine. It includes an introductory essay on the history and social context of toolmaking in Maine; an annotated list of Maine toolmakers; a bibliography of sources of information on Maine toolmakers; and appendices on shipbuilding in Maine, the metallurgy of edge tools in the museum collection, woodworking tools of the 17th and 18th centuries, and a listing of important New England and Canadian edge toolmakers working outside of Maine. This registry is available on the Davistown Museum website and can be accessed by those wishing to research the history of Maine tools in their possession. The author greatly appreciates receiving information about as yet undocumented Maine toolmakers working before 1900.

- Volume 11: *Handbook for Ironmongers: A Glossary of Ferrous Metallurgy Terms* provides definitions pertinent to the survey of the history of ferrous metallurgy in the preceding five volumes of the *Hand Tools in History* series. The glossary defines terminology relevant to the origins and history of ferrous metallurgy, ranging from ancient metallurgical techniques to the later developments in iron and steel production in America. It also contains definitions of modern steelmaking techniques and recent research on topics such as powdered metallurgy, high resolution electron microscopy, and superplasticity. It also defines terms pertaining to the growth and uncontrolled emissions of a pyrotechnic society that manufactured the hand tools that built the machines that now produce biomass-derived consumer products and their toxic chemical byproducts. It is followed by relevant appendices, a bibliography listing sources used to compile this glossary, and a general bibliography on metallurgy. The author also acknowledges and discusses issues of language and the interpretation of terminology used by ironworkers over a period of centuries. A compilation of the many definitions related to iron and steel and their changing meanings is an important component of our survey of the history of the steel- and toolmaking strategies and techniques and the relationship of these traditions to the accomplishments of New England shipsmiths and their offspring, the edge toolmakers who made shipbuilding tools.

The *Hand Tools in History* series is an ongoing project; new information, citations, and definitions are constantly being added as they are discovered or brought to the author's attention. These updates are posted weekly on the museum website and will appear in future editions. All volumes in the *Hand Tools in History* series are or will soon be available in hard copy editions. All other volumes are or will soon be available as bound soft cover editions, are for sale at The Davistown Museum, Liberty Tool Co., local bookstores and museums, or by order from Amazon.com and BookSurge affiliated bookstores.

Table of Contents

Introduction

The *Registry of Maine Toolmakers* is an ongoing documentation of the toolmakers of Maine's maritime era, 1607 - 1900. Its focus is the edge toolmakers and planemakers who supplied the tools for Maine's ship carpenters and timber harvesters, including blacksmiths who specialized in toolmaking. A discussion of their historical milieu begins with the role of imported English edge tools in the toolkits of the first colonial settlers and a description of the prosperous colonial ironworking, planemaking, and shipbuilding industries that arose in southern New England in the 18th century. In the nineteenth century, a vigorous Maine edge toolmaking industry developed, competing with the larger, more well known edge toolmakers of southern New England. This registry also includes listings of known toolmakers for other trades, including coopers, carriage-makers, cobblers, farriers, gunsmiths, and others.

Information sources used in the compilation of this registry include the labyrinth of the Early American Industry Association's *Directory of American Toolmakers*, "DATM" (Nelson 1999), *A Guide to the Makers of American Wooden Planes*, revised by Thomas L. Elliott (Pollak 2001), *Maine Business Directories* of the 19th century, and *Axe Makers of Maine* (Yeaton 2000). Previously unknown toolmakers have also been identified during the last 38 years as a result of the author's search for early tools for the Jonesport Wood Company (Liberty Tool Co.). Other sources include tools found and donated by the Liberty Tool Company customers, benefactors of the Davistown Museum, and the Bob Jones collection of Maine-made planes. The *Maine Business Directories* provide a particularly rich source of information on edge and other tool manufacturers who worked during the years in which the directories were published.

The 1855 *Maine Business Directory* lists over 950 blacksmiths working in Maine. Almost all community blacksmiths (if not shipsmiths or other specific tradespeople, such as farrier or shovel-maker) would have been occasional, if not frequent, makers of one-of-a-kind hand tools for tool users in the communities they served. Aside from making oxen shoes, much of the work of early community blacksmiths would also have involved the repair or reforging of hand tools. The collection of tools in the Davistown Museum includes many examples of worn out steel rasps, which have been reforged into other useful implements. These and most other one-of-a-kind forged hand tools were not usually signed by their makers, but they constitute compelling evidence of the diversity of implements made by community blacksmiths before the era of trade specialization (after 1840) eroded their role as community toolmakers. Many of these blacksmiths would have made only small quantities of edge tools for Maine's flourishing shipbuilding industry. As manufacturers of unsigned natural, forged, or weld steel edge tools (see Appendix C for definitions), their identities will often remain forever obscure. Only a few of Maine's many thousands of 18th and 19th century working blacksmiths are listed in this registry. After 1900, this *Registry* includes only contemporary planemakers working in Maine in the late 20th century.

By constructing this registry, we hope to draw attention to the need for further research to identify the early toolmakers working in the Province of Maine before 1820 and the many Maine toolmakers working after 1820 who played such an important role in the rise of Maine's shipbuilding industry. We know very little about who the toolmakers working before 1820 were, where they lived, what tools they made, and why they made the tools they did. We can be thankful that one major artifact of colonial era material culture, the Merchant wantage rule, survives from this earlier time and is now in the Davistown Museum collection. We hope this spectacular fragment of American history and the *Registry of Maine Toolmakers* that it inspired will be the jumping off point for further research into the identities and activities of the many Maine toolmakers who played such an important role in the economy of the Province of Maine and the vigorous shipbuilding industry that followed after the Province of Maine became a state in 1820.

Woodworking tools predominated in the toolkits of the earliest settlers of the Province of Maine. Despite the obvious role of imported tools from England, house wrights, millwrights, shipbuilders, coopers, and timber harvesters were also dependent on the hand tools produced by the many blacksmiths who accompanied the great migration to New England and their descendants. One mission of this registry is the identification of the edge toolmakers, blacksmiths, and planemakers who manufactured the tools in those first kits. The hand-made hand tools of the 17th and 18th century, many imported from England, but many forged in America, gradually gave way to factory-made hand tools in the 19th century. After the Civil War, the availability of efficient die-forging machinery using low carbon steel made mass production of factory-made tools possible, gradually overtaking the production of hand-forged tools by individual blacksmiths and small factories. Nevertheless, many small independent tool manufacturing companies in Maine and New England flourished in the brief interregnum between 1820 and 1885, when an amazing growth took place in New England and Maine's manufacturing economy. This growth helped sustain Maine's golden age of shipbuilding, providing many of the cargoes (lumber, lime, ice, cotton, and coal) on its coasting schooners and transoceanic Downeasters. By 1885, many of these toolmakers, especially Maine's edge toolmakers, had disappeared or faded into obscurity, as a small number of large tool manufacturing companies came to dominate hand tool production in the late 19th century.

Out of almost 1,000 toolmakers listed in this registry, less than 100 are known to have worked before 1840. Only six toolmakers can be identified as working during the 17th or 18th century in the Province of Maine. Of these, four are represented by tools in the collection of the Davistown Museum (Joseph Metcalf, Thomas Waterman, John Flyn and

Robert Merchant). The museum has no examples from the other early Maine toolmakers, such as Samuel Dennett (chisels and caulking irons) or John Brown of Pemaquid, Maine's first known blacksmith (died 1659). Hundreds of as yet unidentified toolmakers also worked in Maine during this time. Early 19th century Maine toolmakers are also difficult to locate and document. We have identified only a small portion of the hundreds of toolmakers who worked in the first two decades of the 19th century. It is hoped that additional 18th century and early 19th century Maine toolmakers will be identified as Maine tool collectors and historians become aware of the existence of this registry and as we continue the search for New England's early tools and toolmakers.

Maine toolmakers working before 1840: A preliminary listing

Richard Bailey
Calvin Barstow
Flint Barton
Wilibaldus Castner
Joseph Batchelder
I. Batchelder
Samuel Benjamin
T. Bliss
William Breck
Issac Butterfield Jr.
James Bowker Sr.
William Chase
Benjamin Church
Nehemiah Curtis
Isaiah and Jonathan Harding Crooker

Andrew Dunning
Joseph Dustin
Donald Fernald
Stephen Grant Sr.
John Gilmore Sr.
Elisha and Howland Hatch
Samuel Hills
Ezechial and Aaron Hinckley
A. Harmon
George Hight
Caleb Howard
James Jones
J.C. Jewett
William and Richard Mayberry
James McFarland
Noah Mead

Bradley Mowry
James Nealley
David Nichols
A. Parris
John Patten
S. Sewall
Colonel William Stanwood
Theodore Stone
B. Todd
Levi Towle
Joshua Treat
Charles Turner
Nathan Woodward
Js. Wright
Th. Waugh
John Whorff

European Precedents

The motivation for the compilation of the *Registry of Maine Toolmakers* is, in part, a result of almost forty years of the search for woodworking tools by the Jonesport Wood Company (Liberty Tool Co.) for its many customers. Out of the thousands of planes and edge tools recovered and sold by the Liberty Tool Co., several hundred significant specimens, often signed by their makers, are in the collection of the Davistown Museum. Numerous generous donors have added other important tools to the collection.

The Davistown Museum exhibition "An Archaeology of Tools" ends with the classic period of American toolmaking, which culminated in the early 20th century (1930), and the rapid expansion in the variety of iron and steel alloys, manufacturing processes, and tool designs that characterize this industrial florescence. The further back we go when we examine earlier periods of American technological history, the more we find tools that were imported from Europe, either by immigrants or commercial trading companies, in early settlers' toolkits. This raises questions about the origins of early tools in our and other collections that were found in New England, not only in archaeological sites but also in workshops, cellars, old factories, and other industrial environments, and that are, in many cases, one or two centuries old.

When the first settlers arrived in coastal Maine in 1607 to attempt settlement at Fort St. George at Popham Point, the indirect process of iron production utilizing blast furnaces had spread through Europe, producing cast iron that needed refining before being forged into edge tools. Smaller direct process bloomeries also smelted iron ore which could be forged into tools or further refined to eliminate slag inclusions. In 1600, European toolmakers were just making the transition from centuries-old direct process natural steel, forged, and indirect process cast iron-derived steelmaking strategies to more modern practices. In particular, the cementation furnace for producing "blister" steel from refined wrought iron was just coming into wide use in England, dominating steel production there after 1685. The older tradition of making steel from partially decarburized cast iron continued to dominate steel-producing strategies in continental Europe into the early 19th century. The steel produced in the cementation furnace, as well as from refined cast iron (German steel), was then welded onto iron handles, sockets, and other tool parts to make edge tools of a higher quality than those forged directly from a loup of smelted iron. "Weld" steel edge tools were the most common form of edge tools produced in America between 1620 and 1860.

Until the mid-18th century, changes in how tools were forged were very gradual, with slow improvement in furnace and forge design and the thermal and mechanical treatment of iron and steel. In the 16th and 17th centuries, southern Germany (Nuremberg,

Augsburg) was the center of flourishing hand tool, sword, armor, and clock-making industries. German steelmakers had access to high quality iron ore containing manganese mined in Styria (Austria). The manganese in this ore helped neutralize the deleterious effects of sulfur, promoting uniform distribution of carbon. This facilitated direct process production of natural steel in Germany's large high shaft "Stucköfen" furnaces, which were almost as large as blast furnaces. German toolmakers were considered the best in Europe. The cutlers of Sheffield, England, and Swedish ax-makers were also highly regarded for the excellence of their edge tools. Nonetheless, tools of English manufacture, rather than German, dominated the toolkits of Maine's first settlers.

Tools that predate the rise of American manufacturers can be found in the Maine State Museum in a small display of iron tools recovered from the Norridgewock, Maine, site of Father Sebastian Rale's mission settlement, last inhabited by remnants of Maine's Abenaki (Wabanaki) Indians. English colonists destroyed this settlement in 1726. The iron tools discovered there in the 20th century were almost certainly made in France and brought to this country with other supplies for the mission in Norridgewock. The Abbe Museum in Bar Harbor also has a particularly important tomahawk that dates to the late 17th century. One of many tomahawks allegedly supplied to Native American fighters during the Indian Wars and in excellent condition, the Bonaventure hatchet from the Frank Siebert collection (Fig. 1) illustrates the finesse of the European blacksmith in the era preceding weld steel and cast steel edge tools. These archaeological remnants are a reminder that Maine's first European settlers, whether French or English, brought the majority of their tools and iron implements with them when they immigrated to North America. Another excellent example of an early imported edge tool, almost certainly used for shipbuilding is the "Jonesport Broad Ax," located in the collection of the Wilson Museum in Castine, Maine (Fig. 2). This broad ax was discovered sometime before 1921 during a cellar excavation and presented to the museum by George Cousins in 1921. The style of this ax is distinctly European (probably English) and characterized by the light weight poll typical of 17th and early 18th century axes brought to this country by the first settlers. Of additional interest is the maker's hallmark, which consists of

Figure 1 Abbe Museum Bonaventure Hatchet, Frank T. Siebert Collection, French manufacture, 1695, formerly owned by Lewis Lolar, Penobscot. Photo by Stephen Bicknell.

Figure 2 Jonesport Ax, Wilson Museum, Castine, Maine, 12 ½" long, 11" wide, 5" wide poll, 7 lbs. 13 oz, c. 1650-1700. This ax has a maker's hallmark, possibly a touchmark associated with a guild.

initials stamped in the center of the ax, suggesting a possible connection with the burgeoning English market economy that arose in late Elizabethan times. Only a few edge tools and planes dating before 1700 have manufacturer's signatures or hallmarks.

American Iron

The majority of the edge tools used in the colonies in early 17th century were European in origin and similar to the Jonesport broad ax or the Abbe Museum Bonaventure tomahawk in their ferrous metallurgy, either natural steel, forged steel from refined wrought iron, or German steel made from decarburized cast iron. The first indication of the end of colonial America's early dependence on English-made edge tools was the construction and operation of the Saugus Ironworks (near Lynn, MA), 1646. The Saugus blast furnace, refinery, chafery, and blacksmith shop were as advanced in design as any in Europe. This facility illustrated the colonists' early capability of producing significant quantities of cast iron, wrought iron, and blacksmith-made forged steel edge tools especially suited for the needs of an economy dependent on ship carpenters and timber harvesters. The Saugus Ironworks failed after its first few decades of operation due to financial mismanagement, internal labor disputes, and a lack of easily accessible bog iron ore.

Other bog iron bloomeries were soon established at Hanover (1650?), Taunton (1652), Dartmouth (1656), Rowley (1668), and many other southern New England locations. Late in the 17th century, American colonists also began mining New England's most important ore field near Salisbury, Connecticut. The establishment of the many forges in bog iron rich southeastern Massachusetts marks the expansion of a colonial iron industry that by 1720, was providing refined bog iron bar stock of sufficient quality to enable American blacksmiths to compete with their English counterparts. Only recently has the colonial bog iron industry become of interest as a late 17th century alternative source of iron for colonial smiths, a topic explored in volume 7 of the *Hand Tools in History* series *Art of the Edge Tool* (Brack 2008a).

Coastal Maine's earliest identified blacksmith, John Brown of Pemaquid, would almost certainly have made hand tools while living at Pemaquid. Imported English and Swedish iron bar stock would have been his only source of quality malleable iron before 1646, unless he dried and smelted local bog iron. After the establishment of the Saugus Ironworks and other forges in southeastern New England, domestically produced iron, perhaps of lesser quality than the imported iron, would soon have comprised a cargo on New England's coasting schooners.

Changing Toolmaking Technologies

The principal toolkit of Maine's first settlers, and one that dominated commercial activities until the rise of the Industrial Revolution in the 19th century, consisted of a basic group of woodworking tools: adz, broadax, felling and hewing axes, drawknife, auger, froe, pitsaw, and hand plane. Supplementing these tools in most toolkits were hatchets, handsaws, mauls, hammers, scrapers, squares, and other measuring devices, such as marking gauges and travelers, and a few basic agricultural implements (mattocks, shovel, hoe) and the ever-essential hunting knife. These basic tools were the key to the survival of the first settlers coming to New England and formed the nucleus of any established shipyard or woodworker's workshop for over 250 years. Other ancillary trades were also associated with the activities of New England's early woodworkers, such as blacksmith, cobbler, carriage-maker, cooper, coffin-maker, and associated domestic vocations, such as flax-dresser, candle-maker, weaver, and others. In many cases, particularly in the early colonial period, the shipwright and his family were also their own blacksmiths, cobblers, coopers, and flax-dressers. Early trade specialization was the first indication of a growing market economy. Maine's first toolmakers were the farmers, millwrights, and shipwrights forced to supplement the tools that they brought from England and Europe with crude implements hammered out from bog iron blooms. Since they were not making tools for a market economy in which their signature advertised their wares, few, if any, English or American artisans signed the tools they made for themselves, and these toolmakers remain unidentified. The *Registry of Maine Toolmakers* attempts to identify the myriad individual toolmakers living in Maine who gradually began making tools not only for themselves but also for other nearby artisans and communities. The signatures they put on these tools signal the emergence of a market economy, the key component of the vigorous coasting trade that flourished in New England in the years before the spread of the railroad and steamboat.

The commonly encountered English tools of the 18th and 19th centuries obscure the fact that a robust natural and forged steel edge tool manufacturing industry arose in the colonies beginning in the second decade of the 18th century. Bloomery forges produced loups of wrought iron by the direct process of iron production. These iron blooms were frequently refined into malleable iron bar stock before being made into tools. Blast furnaces produced even larger quantities of cast iron for refinement into malleable iron bar stock by the indirect process of iron production. This iron bar stock was hammered and forged into edge tools, often one at a time, by Maine blacksmiths using hand tools and small water-powered trip hammers. The availability of refined, high quality wrought and malleable iron represented an advance over the direct process slag-containing tools produced directly from bog iron without any subsequent refining. In either case, edge tool production utilizing steel bar stock to "steel" the cutting edges of these iron tools gradually replaced the tedious forging of malleable iron by carburization in the forge fire.

The question of the source of steel bar stock utilized before 1720 remains unresolved. After 1720, domestically-produced steel bar stock became an alternative source for New England's edge toolmakers, all now living west of Wells, Maine, because of the great colonial diaspora of Maine's residents due to the King Philip's War (1676). The forging of primitive bog iron tools at isolated farms and smithies became much less common after 1750. Gordon (1996) notes, by the late 18th century, the colonies were exporting bar iron to England and were the third largest producers of iron in the world. Lack of documentation prevents an accurate assessment of the degree to which tools made by colonial blacksmiths and edge toolmakers supplanted imported English edge tools during this period.

While toolmakers in Maine and New England were forging their handmade tools in isolated water-powered bloomery forges and blacksmith shops, an Industrial Revolution that would forever alter the art of tool making was occurring in England. The invention of a new strategy for manufacturing crucible steel by Benjamin Huntsman in 1742 provided England with a monopoly on the production of high quality "cast steel" for edge tool production for the next century. The introduction of the reverbatory furnace for "puddling" wrought iron in 1784 by Henry Cort, combined with the adaptation of the steam engine to blast furnace operation, further increased the efficiency and productivity of English iron production. Herein lies the roots of the illusion that all edge tools were imported from England. They were not. Maine and New England toolmakers soon adopted the new European advances in blast furnace and puddling furnace design and operation, as well as English innovations in machine tool design and production. During the 18th century, New England toolmakers took advantage of multiple sources of steel, including imported English cast and sheaf steel, to establish a vigorous edge toolmaking industry that supplied the larger felling, timber framing, and shipbuilding tools for the booming shipbuilding industry.

Toolmakers in Maine

The Davistown Museum tool collection exhibition consists of a mixture of smaller English cast steel edge tools, especially carving tools and plane blades, American-made natural, forged, and weld steel tools, and imported weld cast steel tools. By the late 18th century, American blacksmiths were producing many of the larger heavy duty natural, forged, and weld steel timber framing and harvesting tools (slicks, broad axes, adzes, and augers) so important to the shipbuilders of the Gulf of Maine region. After 1780, discs of English cast steel were imported to America and incorporated in American-made edge tools of equal quality to tools imported from England.

The most important sources of information on the predominance of smaller English joining and carving tools are Goodman's (1964, 1968) *The History of Woodworking Tools* and his classic study, *British Planemakers from 1770*. The latter contains a comprehensive listing of English planemakers, specimens of which often turn up in American tool chests, and an extensive catalog of British plane blade manufacturers. Though American planemaking began in the late 17th century in southern New England, reliance on English cast steel plane blades continued almost to the Civil War. The importance of the Sheffield, Birmingham, and Lancashire steel industries and their major tool designers, manufacturers, and vendors (Peter Stubs, James Cam, Butcher, Timmins, etc.) is demonstrated repeatedly by the frequent appearance of English-made hand tools in American industries, crafts, and tool collections dating before 1860.

By the early 18th century, iron ore deposits to the south of New England became an important source of American iron, gradually supplanting foreign imports for all but the best grades of refined iron and steel. Coastal New York, New Jersey, and Maryland were all sources of colonial iron. Due to large iron ore deposits in eastern Pennsylvania and in the Pittsburgh area, Pennsylvania soon became the most important iron-producing region and remained so until the mid-19th century, when ore fields in Michigan and Minnesota were opened. Coasting traders began bringing iron from Pennsylvania, New York, and Maryland to New England by 1720 but not before a vigorous colonial bog iron mining industry had been established in southeastern New England. The Saugus ironworks, combined with southern New England's bog iron forges, provided the opportunity for many English blacksmiths who had arrived in the great migration (1630 - 1645) to learn the trade of toolmaking. Some of these blacksmiths, such as John Brown, settled in the ribbon of villages along Maine's coast in the 17th century.

Maine's first colonial settlers had several options for supplying their need for edge tools. Other than food, nothing was more essential for survival than steel edge tools and iron agricultural implements. Isolated farmers and fishermen needing an iron tool could utilize bog iron, common everywhere in the swamps of Atlantic coastal plain, dry it, forge it into a bloom in an open hearth furnace, and make the primitive slag-laced tools still occasionally found today in tool chests and collections. A second option would have been Swedish iron or English edge tools available at local trading posts (Cushnoc, Pemaquid, Pejepscot), village wharfs, or from coasting traders, which were an expensive commodity many settlers could not afford. In between these two stark alternatives lay a solution that helps explain the successful survival of isolated colonial settlements in coastal Maine and elsewhere. Once refined bog iron bar stock made in the colonies was available as an alternative to imported iron and edge tools, American blacksmiths hammered out significant numbers of hand-forged tools for the use of communities where they lived and worked. The proof of this lies in the large numbers of forged iron implements, hardware, and natural or forged steel edge tools that frequently appear in New England tool chests. The rather primitive appearing forged chisel, auger, or drawknife made in colonial New England in 1690 is difficult to distinguish from one made in 1760. Imported English tools, in contrast, have distinctive stylistic characteristics (handles, forms), often have the hallmark of their makers, and look a little more sophisticated than their colonial counterparts. Both imported English and colonial-made hand tools would be found in 18th century Maine toolkits.

The Indian Wars that started in 1676 and resulted in the depopulation of coastal Maine halted most tool production in Maine between 1676 and 1740. The Merchant wantage rule in the Davistown Museum collection, an anomaly of this interregnum, illustrates that even in an isolated Maine village in 1720 in the middle of the French and Indian Wars, colonial toolmakers could produce an artifact as finely crafted as any European import.

By the time of the American Revolution, colonial blacksmiths were capable of an amazing production of weapons and iron equipment best symbolized by the huge chain which was manufactured to block the British from entering the Hudson River. After the Revolution, importation of English tools continued. Helped by the trade embargo (1807) and the War of 1812, a vigorous American tool manufacturing industry arose, leading to the industrial explosion of the 1840s. The small foundries and toolmaking enterprises of the early 19th century utilized both imported weld (blister) and English crucible cast steel and American-made blister steel in edge tool production, dominating the manufacture of heavy duty framing tools for shipbuilding and timber-harvesting. Imported English cast steel plane blades, carving tools, and saw steel illustrated the concurrent florescence of a Sheffield edge tool industry whose reputation, at least for these applications, was not superceded until American toolmakers mastered the alchemy of crucible cast steel production during and after the Civil War.

The anomaly of numerous high quality steel edge tools in the Davistown Museum collection that were made by Maine, New England, or unidentified early 19th century makers and not stamped "cast steel" suggests that enterprising American blacksmiths and small forges were capable of producing steel edge tools of equal quality to the best examples of imported English cast steel prior to their mastery of England's crucible steel production methods (c. 1860). These high quality steel edge tools may have been made from "sheaf" steel, i.e. reworked blister steel refined again by knowledgeable American blacksmiths specifically for the New England shipbuilding and timber-harvesting industries. Even more intriguing, they may have been produced by esoteric methods now almost forgotten. For example, the halting of the fining of cast iron in a puddling furnace while there was still sufficient carbon content to form steel. Another example is co-fusion, the carburizing of wrought iron submerged in liquid cast iron. A third example is the reversing of the fining process in an open hearth furnace by adding carbon to malleable iron to produce steel. These observations apply especially to the larger tools of the shipwright, such as slicks, broad axes, framing chisels, and lipped adzes, not commonly encountered with English hallmarks. The art and science of the ferrous metallurgy of these 18th and early 19th century Maine and New England edge toolmakers remains undocumented.

Thus the mystery of the history of toolmaking in the Province of Maine before 1820. In 1607 - 1676, its first colonial dominion, Maine shared the challenges, shortages, resources, and traditions of southern New England toolmakers. In Maine, the Indian Wars interrupted both the settlement and evolution of commercial toolmaking for decades. When gradual resettlement after 1720 became the land rush of 1767 - 1820, imported English tools and tools made in southern New England were gradually supplanted by the rise of an as yet undocumented Maine toolmaking industry. The purpose of this registry is to further identify Maine toolmakers working during these years and during the boomtown years of Maine's shipbuilding era, 1820 - 1885. Maine's vigorous edge toolmaking industry briefly overlapped with the factory system of mass production of edge tools and other hand tools that evolved after 1860. Once the American factory system was firmly established, English toolmakers became less important as a source of hand tools. Only a few decades remained before most Maine edge toolmaking would be superseded by the mass production of southern New England and New York toolmaking factories. Ironically, Maine edge toolmakers, who often made their fine edge tools out of imported English sheaf and/or cast steel, may have mastered alternative strategies for making edge tools at the same time as the Buck Brothers, Underhills, T. H. Witherby, and others learned the secret of cast steel production.

For a few decades after the Civil War, Maine's edge toolmakers continued making tools for both Maine shipwrights and timber harvesters, but the factory system of mass tool production was growing rapidly in the second half of the 19th century. Low carbon steel, produced first by the Bessemer process and then by the Siemens-Martin open hearth process, made possible the production of huge quantities of inexpensive die-forged tools made in Massachusetts and Connecticut that replaced most Maine toolmaking operations.

Only high quality edge tools, which could not be made with low carbon steel or malleable iron, continued to be made in Maine's smaller factories. Ax-making was the last Maine tool industry to be subsumed by the massive American conglomerates, such as the Stanley Tool Company, which dominated hand tool production until the mid-20th century. The American Axe & Tool Co.'s purchase of many Maine ax-makers in 1889 signaled the end of the era when individual craftsmen made their own tools, either by themselves or in the small forges and factories that dotted the Maine landscape in the 19th century. The American Axe & Tool Co. eventually disintegrated and many Maine edge toolmakers, such as Emerson & Stevens Co. and Snow & Nealley, continued operation well into the 20th century, utilizing the modern die forging techniques first pioneered at the Collins & Co. ax factory in Connecticut in the 1840s. The hundreds of ax-makers working before 1889 became just a handful of companies in the 20th century. But what was the principal use of the edge tools made by Maine and New England toolmakers during the previous 300 years?

Shipbuilding in Maine 1607 - 1900

The most important consideration in documenting Maine's toolmakers working before 1900 is the fact that shipbuilding formed the nucleus of Maine's economy until the Civil War and remained an important industry even into the early 20th century. The florescence of Maine's resource-derived activities of fishing and timber-harvesting would have had no viability without the ships that transported these and other commodities. New England's ship designs, shipbuilders, and their cargoes, routes, captains, and adventures are well documented. In contrast, the toolmakers who made the hand tools that made New England's vigorous maritime economy possible have been overlooked or forgotten.

Our search for early Maine toolmakers includes artisans, both blacksmiths and shipsmiths, who made tools for the shipwright, as well as the timber harvester. Important subsidiary categories of tool users include joiners doing finish work on ships, the all important cooper, who provided so much of the cargoes of the coasting and West Indies trades, and ancillary occupations of importance, such as spar-maker, block-maker, rope-maker, sail-maker, wagon-maker, and supporting domestic trades, all of whom needed hand tools for their work. With the blossoming of the Industrial Revolution and the appearance of machine-made tools and power-driven woodworking tools in Maine shipyards (1840-1865), the complexities of shipbuilding toolkits increased and included both patternmakers' and machinists' tools.

The roots of several centuries of a vigorous state of Maine toolmaking industry lie far in the past. The first ship known to have been constructed in Maine by EuroAmericans was the pinnace *Virginia* built at Fort Popham in 1607, which later sailed back to England with some of the discouraged Popham settlers. The turmoil of the English revolution in the 1640s reduced the supply and availability of English ships, and New England colonists began a long tradition of shipbuilding that finally ended with the era of the racing Gloucester fishermen and, in Maine, with the huge four-, five- and six-masted schooners built at Bath, Waldoboro, and other Maine locations in the later years of the 19th and first decades of the 20th centuries. Between this final era of bulk cargo schooners and the first cod fishermen of colonial times lies a remarkable florescence of shipbuilding that culminated in Maine in the golden age of shipbuilding between the years 1820 and 1885. The Chesapeake Bay region, New York, and southern New England were the centers of American shipbuilding activity until the early years of the 19th century. With the exception of Kittery and York, few ships were built in Maine until the end of the turmoil of the French and Indian Wars (1686 - 1759). After the Treaty of Paris in 1763 and the opening of eastern Maine, including the Pleasant River settlements, a vast influx of immigrants began moving to Maine, and vigorous, though often undocumented, shipbuilding activities began. After the Panic of 1857 and the Civil War, Maine's shipbuilding continued to flourish despite the opening of the Suez Canal (1867) and the growing use of steam power for ocean-going and coasting passenger carriers. The longevity and durability of this shipbuilding era are testaments to the skills of the shipbuilders and sailors of the era and to the skilled toolmakers who made their shipbuilding tools.

The Gulf of Maine bioregion includes important Massachusetts shipbuilding centers north of Cape Cod, i.e. North River (Scituate, MA), East Boston (MA), Mystic River (Medford, Salem, and Essex, MA), the lower Merrimac River area (Newburyport, MA), and the Portsmouth (NH) tidewater of the Piscataqua River. These centers dominated shipbuilding activities in New England until the early 19th century. Timber-harvesting, privateering, the growing cod fishery, the neutral trade, and a vigorous West Indies trade prompted increasing shipbuilding activities in Maine until, by the 1820s, Maine was approaching the importance of New York and Massachusetts as a shipbuilding center. By the 1830s, Maine had equaled the output in number, though not in size, of those locations and, by the 1840s, Maine led the nation in tonnage and number of ships built.

Stahl (1956) notes that, by 1838, the Waldoboro customs district (Thomaston, Warren, Waldoboro, and Boothbay) was surpassed in ownership of sailing vessels in Maine only by Portland (44,661 tons vs. 56,191). Between 1842 and 1856 the Waldoboro district surpassed Portland, with owned tonnage peaking in 1856 at 155,783 tons. This florescence of shipbuilding in coastal Maine, just down the St. Georges and Medomak rivers from Liberty and Montville, closely correlates with the rise of a vigorous edge tool manufacturing industry in Maine, New Hampshire, and Massachusetts. The booming years of shipbuilding in the Waldoboro customs district also coincides with the high point in population levels of the water mill towns of Liberty Village, Montville, Searsmont, South Liberty, and other nearby communities.

While the most prolific makers of shipbuilders' edge tools for Maine's shipyards in the 1830s and 1840s were the Underhill clan of Nashua and Boston (see Appendix G), Maine's growing number of shipyards resulted in a vigorous edge tool manufacturing industry within its borders. Though examples of southern New England, New Brunswick, Canada, and New Hampshire edge tool- and planemakers dominate the toolkits of Maine's shipbuilders and cannot be ignored in the study of the hegemony of shipbuilding in Maine in the years after 1840, the rise of a vigorous edge tool- and planemaking industry in Maine after 1820 is an intriguing and important component of Maine's state history. The availability of high quality, inexpensive, New England-made edge tools, Maine's skilled shipbuilding community, and its deep water harbors and rivers were the key components of the success of the shipbuilding industry of Maine in the 19th century. In turn, Maine's superbly constructed sailing vessels were inexpensive to build and man and constituted the majority of America's coasting and ocean-going fleet after 1840.

For a chronological sketch of the history of the shipbuilding industry in Maine, see the History of Shipbuilding in Maine in Appendix B.

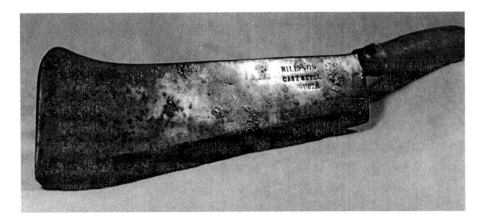

Figure 3 Cleaver made by Billings of Augusta, one of the many edge toolmakers working in the Kennebec River watershed in the 19th century. Cast steel, 15 5/8" long, 10 ¼" blade (ID # 61601T1).

Maine's Shipsmiths and Edge Toolmakers

In the Province of Maine, the early years of the republic were characterized by a booming economy based upon forest resources, fishing, and the flourishing foreign trade, which was suddenly cut off by the trade embargo of 1807 and the War of 1812. After the depression in the second decade of the 19th century, the booming fishing industry documented by O'Leary (1996) in *Maine Sea Fisheries* supplanted the foreign trade as Maine's most important maritime activity. The combination of vigorous fishing, forestry, and lime production industries gave rise not only to the planemakers of coastal Maine but also to a robust Kennebec River watershed edge tool manufacturing industry, the origins of which are not clearly understood.

One of the more obscure facets of the now almost forgotten history of edge toolmaking is the role of the shipsmith as edge toolmaker before the modern era (>1840). The trades of knife and sword cutler, historically the most important forms of edge tools, are poorly documented and not yet the subject of any comprehensive study. From the early Iron Age to the early modern era, most edge tools used by the shipwright were forged by the same blacksmith who made the iron fittings that all wooden ships required as part of their construction. This observation applies to ships built to transport Roman Legions across the Mediterranean or from France to England, to English ships built in the Thames River, or to the early shipbuilders of Maine and colonial New England. The shipsmith was the blacksmith who lived in shipbuilding communities and forged iron fittings, such as spikes, futtock bolts, deck and knee irons, side rings, and other hardware needed on every wooden ship. In many cases, the shipsmith also forged the edge tools needed by the shipwright, often one-of-a-kind edge tools of a specific size or shape, which could not simply be ordered from a Lancashire or Sheffield edge toolmaker. Early colonial blacksmiths who were multitasking shipsmiths and edge toolmakers were permanent residents of larger New England and Maine shipbuilding communities, especially in the 18th century, as ports, such as Boston, Salem, and Scituate, grew and prospered. Initially, for some isolated communities, such as those in downeast Maine, many shipsmiths would have moved from community to community, either bringing their small forges with them on coasting schooners, which were their only transportation, or utilizing equipment and small furnaces at existing shipyards. The iron and steel bar stock essential for their trade was one of the more ubiquitous commodities on coasting vessels, often, in the case of iron bar stock, carried as kentledge (ballast). A more detailed review of New England's shipsmiths and edge toolmakers is contained within volume 8 of The Davistown Museum's *Hand Tools in History* series *Art of the Edge Tool: The Ferrous Metallurgy of New England Shipsmiths and Toolmakers* (Brack 2008b).

While we have evidence of itinerant blacksmiths traveling to and settling in Maine's coastal communities both in the colonial period prior to the great diaspora of 1676 and again beginning in the middle years of the 18th century, there is a gap in our knowledge about edge toolmakers who settled in the far upriver regions of interior Maine, well away from the coastal settlements, in the first five decades of the 19th century. These edge toolmakers, predominantly ax-makers, have left their signed tools for us to find in tool chests, barns, and collections throughout New England. Examples may be seen both at the Davistown Museum and at other New England museums that also have edge tools in their collections. Names such as Whorff, Billings, Lovejoy, Thaxter, Bragg, Graves, and Haskell are typical of Maine-made edge tools in the Davistown Museum collection that have been found in Maine. They signal the presence of a robust community of ax-makers and edge toolmakers who lived far inland, making their tools from bar iron and blister steel. The iron and steel they used must have been brought upriver by the same coastal traders who then returned to the cities of the Atlantic seaboard with the forest products harvested with these edge tools. Maine's only known blast furnace, the Katahdin Iron Works, furnished little or none of the bar iron utilized by these blacksmiths as they hammered out their natural or weld steel edge tools. The rise and fall of these inland ax-makers parallels and, in fact, documents the boomtown years of Maine's inland forest product economy.

In contrast, edge toolmakers, such as Vaughn and Pardoe of Union, Libby and Bolton of Portland, T. C. Jackson of Bath, and Ricker of Cherryfield, occupied coastal locations and their tools help document the booming shipbuilding economy of Maine. While the majority of adzes, slicks, and other edge tools utilized in Maine's shipyards in the 19th century were produced outside of Maine by famous and prolific edge toolmakers like the Underhill clan, Thomas Witherby of CT, the Buck Brothers of Millbury, MA, or Josiah Fowler of St. Johns, New Brunswick, a thriving community of Maine edge toolmakers produced a surprisingly large minority of the edge tools used in Maine's shipyards and forests after the early 19th century. These tools survive with the signatures of individual edge toolmakers whose identities would otherwise have been lost or forgotten.

Coopers and Others

The individuals and small shops that made tools for one of Maine's most important early industries, the cooperage, are among the most difficult of all early toolmakers to document. Coopers played an essential role in Maine and New England's maritime culture by making staves for barrels, trawl-line tubs, lime casks, hogsheads, firkins, and other essential woodenware found on every fishing vessel and coasting packet or trader for over 250 years. As with many other kinds of tools, coopers' tools were often handmade by the farmers and woodsmen who made many of these products during Maine's long winters. As Maine's robust market economy gradually emerged in the late 18th century, individual families specialized in making only coopers' wares, and individual toolmakers made coopers' tools as part of their output of hand tools for sale. Some Maine toolmakers may even have specialized in making only coopers' tools. The town of Liberty, in which the Davistown Museum is located, had over 30 working coopers at the time of the Civil War in the twilight of its boomtown years. The mill towns of Montville and Liberty reached peak population levels in the 1840s before residents began migrating to the rich bottomlands of the Ohio River valley or to California after 1848, lured by the gold rush. This peak in population and water-mill-derived manufacturing coincided closely with the florescence of shipbuilding in central coastal shipyards east of Bath and the Kennebec River in the 1830s - 1850s, including the Waldoboro Customs District. The demand for coopers in Liberty probably lasted longer than in most other coastal locations because the vigorous lime industry in Thomaston required large numbers of lime casks late into the 19th century, long after the factory system and its machinery had rendered the handmade woodenware of most coopers obsolete.

Other 19th century industries that gave rise to specialized toolmaking include agriculture, the ice trade that flourished in Maine in the middle decades of the 19th century, and the granite trade that flourished in the late 19th century. All of these industries utilized edge tools that could have been made by any toolmaker in small quantities. The North Wayne Scythe Co. typifies the rise of the American factory system. A robust agricultural implement manufacturing business, it was among the world's largest producers of scythes during its most active years. Hundreds of manufacturers of agricultural equipment working in the 19th century remain unlisted in this directory simply because our focus is on edge toolmakers and planemakers. Most manufacturers of Maine's ice-harvesting tools remain unidentified. Bicknell Manufacturing Company is a very late 19th century producer of granite quarrying tools, but most earlier blacksmiths and companies making quarrying tools also remain unidentified.

The Origins of Planemaking

A review of European plane designs helps illustrate the changing forms of American hand planes in the 18th and 19th centuries (Fig. 4). These forms help date the planes made after 1790 by Maine's first planemakers (Fig. 9). While the blacksmith fashioning edge tools for shipbuilding remains the most important toolmaker of New England maritime culture, the story of planemakers who made the wooden bodies to hold the plane blades is one of the most interesting chapters of our industrial history. Unlike most other edge tools, almost all plane blades were imported from Sheffield, England, until the mid-19th century, as exemplified by the Waterman plane (Fig. 9). Yet, most hand plane bodies were made by the settlers themselves out of the birch, oak, and, later, beech so ready to hand in New England or from exotic woods derived from the West Indies trades, such as lignum vitae, ironwood, rosewood, and coco bolo.

Goodman (1964) traces the history of planemaking back at least to the Romans. The planes recovered from the wreck of Henry the VIII's *Mary Rose*, which sank in Portsmouth Harbor, England, in 1545, show that continental plane forms dominated the toolkits of 16th century English shipwrights. The "modern" forms of the hand planes of the 19th century derive not from English prototypes, nor even from the medieval, rather French-looking hand planes found on the Mary Rose, but from Dutch prototypes that influenced English and then American plane design. More information can be found in Wing's (n.d.) *The case for Francis Purdew or granfurdeus disputatus*. The basic form of moulding and jointer's planes remains virtually unchanged since the heyday of the Dutch empire in the early 17th century.

A few characteristics differentiate 17th and 18th century American hand planes from the ubiquitous productions of small and large 19th century American workshops as follows:

- The early moulding planes were larger, 10 ¼" to 10 ¾" compared to the shorter 9 or 9 ½" 19th century planes. Production of hand planes of a uniform length began in England by 1780 and was commonplace in the United States by 1810.
- American planes made before 1800 retained the strong beveling along the top edges characteristic of European planes of the 18th century. This chamfering gradually diminished over a short period of time from 1780 - 1820. Nineteenth century moulding planes show little or no beveling on their outer edges.
- Eighteenth century jointer planes, often, but not always, have an offset handle. The modern design of jointer planes, with their closed or pistol grip handles, appeared as early as 1700 in England, more or less at the same time that Moxon ([1703] 1989) published his illustration of medieval planes (Fig. 4), copied from Félbien's illustration in *Principes de l'Architecture* of 1676. The medieval forms illustrated in Moxon seldom appear even in early colonial era toolkits.

- Eighteenth century jointer planes usually have closed pistol grip handles. Smaller 18th century fore and jack planes, in contrast, have a characteristic open grip handle that did not evolve into the closed pistol grip style until the 19th century.

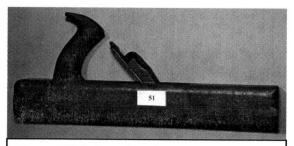

Figure 5 This fore plane has an open grip handle typical of the style used on fore planes during the late 17th century and most of the 18th century. Gutter plane (ID# 111001T-9) in the Davistown Museum MIII collection.

Figure 6 Levi Tinkham fore plane c. 1840. The open grip handle on this panel plane is offset, a typical trait of 18th century planes. From the Davistown Museum MIII collection (ID# TCD1003).

- While yellow birch and occasionally oak were the predominant wood used for making early American planes, beech became the chosen wood for almost all American planemakers at the beginning of the 19th century, with the exception of the tropical woods used by New England's coastal shipbuilders, who did not usually offer the planes they made for sale.

For an excellent and more detailed examination of changing styles of planemaking see Whelan (1993).

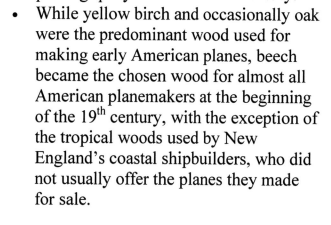

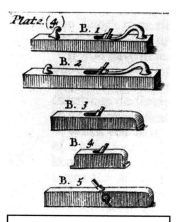

Figure 7 This birch fore plane has a closed pistol grip handle, a design that supplemented the open grip handle on most fore and jack planes during the first few decades of the 19th century.

Figure 4 Moxon's illustrations of medieval era planes with handle designs that pre-date the open grip and pistol grip handles were already obsolete at the time of his third edition. Moxon. [1703] 1989. *Mechanick Exercises or the Doctrine of Handy-works*. Third Edition. Morristown, NJ: The Astragal Press. pg. 68.

Planemaking in Colonial New England

When attempting to identify the earliest planemakers working in Maine, one encounters the inescapable fact that a robust community of planemakers arose in southern New England, centered in the Blackstone River Valley, as early as 1700, well before the Province of Maine became a state in 1820. The most renowned southern New England planemakers were the Nicholson family. Francis Nicholson of Wrentham has long been considered the first planemaker working in colonial America to sign his tools. Pollak (2001) indicates that Nicholson's working dates are 1728 - 1753. Nicholson worked with his son, John, and the first black planemaker, the slave Cesar Chelor. Their planes are among the most sought after by tool collectors. Donald and Ann Wing (2001) have written about this milieu in their essay, "Planemaking in Eighteenth-century America." As they note, Cesar Chelor was set free upon the death of Francis Nicholson; the planes he made have become an icon of American rhykenology.

Figure 8 Adjustable plow plane stamped "I. NICHOLSON LIVING IN WRENTHAM". Pollak (2001) indicates this stamp is characteristic of planes made between 1733 and 1740. This plane was loaned to the Davistown Museum by Bob Wheeler during 2002.

Collector of 18th century tools and patron of the Davistown Museum, Bob Wheeler, indicates that two other planemakers, Ebenezer Seymour and Nathaniel Potter, may have been making planes in New England at the same time as Francis Nicholson (Wheeler 1993). Ebenezer Seymour was born in Connecticut in 1683 at approximately the same time as Francis Nicholson and is believed to have made planes in the Connecticut town which bears his name. Nathaniel Potter was born in 1693 and apparently made planes in Leicester, Massachusetts, at the same time that Francis was making his tools. Other important early plane makers included Thomas Grant and James Stiles of New York City and Samuel Caruthers of Philadelphia (Wing 2001, 118). The long tradition of southern New England planemakers provides the context for the rise of planemaking in Maine. Among the flood of settlers coming into Maine after the Treaty of Paris in 1763, trained planemakers were probably among the skilled woodworkers who provided the labor force for Maine's growing shipbuilding trades. Only a few of Maine's early planemakers can be identified with certainty; others are just names found on surviving planes in Maine. Hundreds of other planemakers, including the shipwrights who made their own planes, often out of tropical woods, will remain forever unidentified.

The robust community of whalecraft manufacturers working in the New Bedford, MA, area in the early 19th century is not to be overlooked as an important component of the community of edge toolmakers working in Maine and southern New England during the booming years of Maine's shipbuilding industry. The production of whalecraft, i.e. the iron and steel tools used to capture and kill whales, such as harpoons, toggles, flensing irons, spades, and bomb-lances, was often accompanied by the production of edge tools. Lytle (1984) lists these toolmakers in his Appendix 1, adapted in this registry as Appendix H. Whalecraft production continued in the New Bedford area into the early years of the 20th century and was an integral part of New England's edge tool manufacturing industry. Its decline coincided with the decline of the whale fishery. A record of the names and touchmarks of these toolmakers helps identify the location of the manufacture of edge tools that might otherwise remain unidentified. Many a New England blacksmith or edge toolmaker may have learned their trade in the heyday of whalecraft production (1830s) and then traveled north to the booming tool factories, foundries, and shipyards of Massachusetts, New Hampshire, and Maine.

Maine's Earliest Planemakers

Maine's first planemakers would have been among the first colonial settlers living in the isolated fishing and trading villages of ancient Pemaquid, Georgetown Island, Pejepscot, and communities to the west. Block planes, jointer planes, and molding planes for shipbuilding would have been among the first to supplant or replace whatever imported English planes the first settlers brought. Recycled farriers' rasps would have been a common source of steel for replacing plane blades for those who could not afford English blades. No identifiable planes survive from the ancient dominion of colonial Maine prior to the King Philip's War (1676) and the great colonial diaspora that followed. In southern New England and Maine in the 17th century, planemaking was a craft-based occupation where individuals made their own planes for their own use. The tradition of owner-made hand tools continued in Maine and New England well into the 19th century, as illustrated by the Abiel Walker hoard in the Davistown Museum collection (see Appendix E.)

A number of intriguing examples of the work of early commercial planemakers living in the Province of Maine have surfaced and are also in the Davistown Museum collections. Joseph Metcalf (born Medway, MA, 1756), the brother of Luther Metcalf (b. Medway, MA, 1765), moved to Hallowell and then to Winthrop, Maine in 1789. Bob Wheeler has identified eight or nine planes marked "J. Metcalf" in large embossed letters as being present in museum or private collections, including a 10 13/16 inch long yellow birch rabbet plane and a rounding plane in the Davistown Museum collection. Rhykenologists note that 18th century planes were usually made of yellow birch, but, as the forests of southern New England were quickly cut over and the land denuded, beech was substituted for yellow birch by most planemakers. Thomas Waterman (b. Waldoboro, ME, circa 1775), who worked in the last years of the 18th century, is another early Maine planemaker (Fig. 9). The Davistown Museum has also recently obtained an 18th century plane from the collection of Ben Blumenberg (courtesy of Bob Wheeler,) which is a complex moulding plane with an 18th century appearance marked "JOHNFLYN". This plane was found in Warren, Maine, and may possibly predate both the Metcalf and Waterman planes in the museum collection.

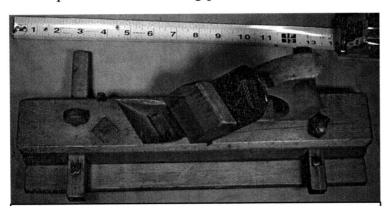

Figure 9 A plane of historic significance in the Davistown Museum collection is the Thomas Waterman panel raising plane with adjustable fence signed "T WATERMAN", with a James Cam cast steel blade. Exquisitely constructed and with distinctly 18th century chamfering, this plane was most certainly made in Maine in the last years of the 18th century and is one of the earliest surviving signed, Maine-made planes.

The trickle of colonial settlers returning to Maine in the first few decades of the 18th century became a flood of new immigrants after the Treaty of Paris at the end of the French and Indian Wars opened the Province of Maine for resettlement (1763). Most immigrants would have brought their own tools, many now made in southern New England, or fashioned their own planes and edge tools as needed. But Maine's rapidly increasing population signaled the dawn of a new era, where sufficient demand existed for both planemakers and edge toolmakers to serve a burgeoning market economy.

While Metcalf, Waterman, Flyn, and other as yet unidentified planemakers were making tools for this emerging market economy, stubborn Maine shipwrights continued to make their own tools for their own use. Thousands of unsigned, owner-made planes still survive in collections and workshops. Of these, only planes made from exotic tropical woods are of interest to collectors. Their origins lie in the long established West Indies trade that began in the 17th century. High quality, durable tropical woods from the Caribbean islands and Bay of Campeche could be obtained easily by New England's shipwrights, who made their own distinctive, but usually unsigned, block and razee planes for the now flourishing shipyards. Emerging commercial planemakers, such as Metcalf and Waterman, didn't normally use tropical woods for their planes (Storer of Brunswick was a notable exception.) It is an interesting unanswered question as to whether their preference for birch and then the ubiquitous beech was a matter of economy, personal preference, hoarding of tropical woods by shipwrights, long established craft tradition inherited from southern New England planemakers, or some other factor. Planes made from tropical wood are almost always unsigned, making them difficult to date, and yet they keep turning up in New England tool collections, especially those associated with Maine shipbuilding communities.

Toolmaking as a Commercial Activity in 19th Century Maine

While Maine's planemakers and shipwrights were making handmade planes based on centuries old craft-based techniques, new methods of toolmaking were evolving in distant locations, such as London and Manchester, England, and then in southern New England. Based upon the invention of new equipment for toolmaking, these changes would gradually alter plane and edge toolmaking techniques in Maine before rendering the hand work in toolmaking nearly obsolete in the late 19th century.

The inventions and innovations of first English, and then American, machine engineers were of hidden, but critical, significance for the evolving historical milieu of Maine toolmakers. The steam engine first evolved in England due to lack of water power. The textile industry, powered by steam engines in England, soon migrated to America with the help of Samuel Slater, who established the first water-powered, mechanized cotton mill in Rhode Island in 1793. The 45 different block-making machines for the Portsmouth, England, naval shipyards, designed by the French inventor Marc Brunel and manufactured by the English engineer, Henry Maudslay, at the Woolwich Arsenal in London between 1802 and 1809, represented the craftsmen's "red sky in the morning." This equipment was still in use as recently as 1950. Slater's mechanized cotton mill and Maudslay's block-making equipment represented the first small wave of an industrial tsunami, which would soon flood England and America with machine tools of every description.

Bench-top lathes, screw- and gear-cutting machines, table engines, compound slide rests, marine steam engines, and many other types of machines that make or power other machinery characterize an Industrial Revolution that developed as a result of innovative English and French engineering. American engineers and designers soon adapted and improved this new machinery. Oliver Evans, Eli Whitney, Simeon North, Elisha Root, and, later, George Corliss are only a few of the American inventors who quickly copied and improved the designs of English engineers to meet the needs of an American factory system, which surpassed England's manufacturing capacity by the mid-19th century. The increased use of machinery to replace hand work by Maine's toolmakers can be noted as early as 1825, when the rotary sawmill replaced not only the more inefficient up and down sawmill, but also the whip (pit) saw and broad ax in many Maine shipyards.

Of particular note is John Hall, the famous Portland, Maine, machinist and gunsmith who improved the design of Simeon North's first milling machine at Harper's Ferry Armory.

> By 1828, John Hall and Simeon North were producing rifles with interchangeable parts... By 1832, the Norths, Hall and the armories had the world's first machine able to "mill" or cut, flat and curved surfaces in iron with a powered, rotary cutter to a

high degree of precision... By the 1840s, anyone could buy a milling machine from the Ames Manufacturing Company of Chicopee, Massachusetts, or the Robbins and Lawrence Company of Windsor, Vermont. (Muir 2000, 129)

The design and implementation of the milling machine and other equipment derived from earlier English models changed toolmaking in Maine from a crafts-based effort, often in scattered, isolated communities, to commercial enterprises centrally located in trading centers such as Portland. No better example of this transition to the factory system of tool production exists than the rise of commercial toolmaking in Bangor, Maine, in the heyday of its boomtown years, 1840 - 1860. By this time, Maine planemakers, such as Benjamin Morrill (1832 – 51) and Percy Rider (1834 – 48) were producing planes in Bangor in much greater numbers than those made by Joseph Metcalf a half century earlier. The availability of new woodworking machinery played a key part in this transition. Even more important in a historical context was the work of some of America's first machinists in the improbable location of Bangor: Samuel Darling and Edmund Bailey, Bangor (1852 – 53), Michael Schwartz (working in Bangor after 1843), and Darling & Schwartz (1854 – 66). The machinist tools they made, often using one-of-a-kind machines that were later perfected and mass-produced, are icons in the history of American hand tool production. They signal the rise of commercial toolmaking in Maine and in America. Only a handful of Darling & Bailey hand tools survive. The Davistown Museum has a Darling & Bailey steel rule in its collection and a Darling & Schwartz depth gauge illustrated in the registry listings.

Other commercial toolmaking centers in Maine include Belfast, which was famous as a center for screwdriver inventors and manufacturers (Issac Allard, F. A. Howard, J. W. Jones), as was Augusta (Zachery Furbish and George Gay). The waterpower of the Clinton, Oakland, and Waterville area gave rise to numerous ax and edge toolmakers (Dunn Edge Tool, Emerson & Stevens, the Billings clan). T. C. Jackson was an important Bath adz maker. George Evans made planes in Norway. Libby & Bolton made edge tools in Portland. E. T. Burrowes, also of Portland, made levels and rules. The names Peavey and Snow & Nealley (Bangor) are icons in the history of American timber-harvesting tools; Snow and Nealley continued to produce high quality axes into the early years of the 21st century. Vaughan & Pardoe of Union made edge tools for shipbuilders in the Waldoboro customs district. All of these toolmakers produced large numbers of tools and were commercially successful precisely because they incorporated much of the machinery of the Industrial Revolution in their tool manufacturing activities.

Of particular interest is an advertisement for the Charles H. Reynolds Co. of Lewiston in the 1855 Maine Business Directory Advertising Supplement. The Reynolds ad indicates that they made both steam engines and boilers. This type of industrial equipment played a key role in gradually eliminating the artisan in many of the activities of toolmaking in the years between 1850 and 1885. Improved designs in trip hammers originated in southern New England in the early 19th century and helped Maine forges, foundries, and edge toolmakers expand their production and serve ever larger communities. While Maine's shipwrights were importing edge tools from the Underhills of New Hampshire and Boston via the coasting trade, Libby & Bolton of Portland were also producing edge tools, some of which were later found in southern New England tool chests. By 1860, the mass production of edge tools by companies like the Collins Axe Co. of Connecticut and Buck Brothers of Milbury, MA, were providing stiff competition for the hundreds of smaller Maine edge toolmakers who flourished from 1820 - 1885. Maine's many ax-makers continued their vigorous production of tools for harvesting Maine's vast forests until the early years of the 20th century, while most other Maine edge toolmakers had faded into obscurity several decades after the end of the Civil War.

By 1865, the mechanized drop-forging of hand tools first utilized in the Collins Axe Factory in Connecticut after 1827 had spread throughout New England. The availability of steam engines to power trip hammers in lieu of water power brought tool factories to cities such as Portland. Cheap, low carbon steel produced by the Bessemer process and then from Siemens-Martin open hearth furnaces helped commercial toolmakers, such as the Diamond Wrench Company of Portland, successfully produce their uniquely designed wrenches. Using similar modern drop-forging techniques, but using higher quality steel from unknown sources, the North Wayne Tool Company produced huge numbers of scythe blades. Russell Phillips of Gardner utilized machinery to produce his gorgeous, and now highly valued, plow planes (1867 - 1870) using recently improved malleable cast iron containing alloy mixes which, even now, remain secret. In the meantime, Maine edge toolmakers were also feeling the impact of the continuous stream of innovations in both machinery and metallurgy for commercial tool production. A vigorous wooden shipbuilding industry continued in Maine coastal communities, especially Bath, but also Damariscotta, Waldoboro, Thomaston, and Belfast, building small schooners for the tenacious coasting trade and huge four- and five-mast schooners to compete with the growing fleet of steam ships and with rail transportation. This was the golden age of American-produced cast steel edge tools, typified by the exquisite slicks and framing chisels of Underhill, Witherby, the Buck Brothers, and a few Maine makers. These tools had sleek smooth surfaces characteristic of martinized steel with its uniform carbon distribution and represent the highest achievement of American toolmakers. The rise of this factory system of tool production helped commercial toolmaking in Maine to flourish briefly. Ironically, many of the measuring tools made in New England factories were used for the manufacture of machinery that soon rendered the use of many edge and other hand tools obsolete.

Maine's first documented Toolmaker: Robert Merchant and his wantage rule

The milieu of the southern New England planemakers provides the historical context for considering the significance of the work of Robert Merchant. His wantage rule is inscribed "Made by Robert Merchant for Noah Emery, Berwick, 1720". Of the many known examples of planes made by Francis or John Nicholson, Ebenezer Seymour, Nathaniel Potter, or Cesar Chelor, who were working in southern New England at this time, or shortly thereafter, none were ever dated. The Merchant wantage rule is, therefore, exceptional, even anomalous, in containing an inscribed date, indicating exactly when it was made and for whom it was made. It is certainly the oldest signed and dated tool presently known to have been made in the Province of Maine. This rule only surfaced very late in the 19th century, having been discovered in a barn in Eliot, Maine and then consigned to an auction in Portsmouth, NH, in 1998. This tool was used for measuring volumes of beer, wine, and other spirits in an era when it was much safer to drink beer from a hogshead than water. This wantage rule is a work of art with its fitted slipcase, meticulous markings, and mellowed hue from years of use measuring volumes of alcoholic beverages. It is also one of the most significant pieces of early American

Figure 10 This carefully constructed rule is signed in script "made by Robert Merchant for Noah Emery" "Berwick" "1720". This signature is clearly visible in the lower photograph. The rule came in a carefully crafted slip case, which requires a ribbon to remove the rule (see photo under the registry listings). No earlier signed measuring tool has come to light in any major public American collection of hand tools (Mercer Museum, Smithsonian, Shelburne Museum, Mystic Seaport, Deerfield, Winterthur, or Sturbridge Village). From The Davistown Museum MI collection (ID# TBW1006).

material culture in any museum or collection in the United States. This rule forms the cornerstone of the tool collection of the Davistown Museum and remains a principal incentive for the search for the identity of other early toolmakers who worked in the Province of Maine.

The discovery of the Merchant wantage rule and its early date of manufacture helped inspire the attempt to locate other early Maine and New England toolmakers. The miracle of its survival for almost 300 years is the source of two observations about the relationship between tools and art, the focus of the collection and exhibitions of the Davistown Museum. First, those rare hand tools that are exceptional in their beauty and excellence of design and craftsmanship are significant as works of art in themselves. As with any sculpture, they give voice to realities about our culture that we cannot put into words. They exist in and of themselves as a significant art form. What they mean or signify can be debated for generations. The second observation is that many hand tools were used to make works of art. The beauty and grace of Maine's ships, from the ubiquitous coasting schooners to Maine's sleek "Downeasters" testify to the art of shipbuilding. Mention the words *"Flying Cloud"* to any knowledgeable Maine maritime history buff and few would argue that this ship, built in Damariscotta in the 1880s, was not a work of art. Many other ships of equal beauty slid down the ways of Maine and New England shipyards. They are an enduring testament to the skills of both the shipwrights who built them and the toolmakers who made their construction possible.

The compilation of the *Registry of Maine Toolmakers* is, in part, an attempt to recognize, document, and advocate the important role Maine toolmakers made in facilitating Maine's robust shipbuilding industry. A significant minority of the tools used by Maine's shipwrights and other woodworking and timber harvesting trades were made in Maine and not in the more well known toolmaking communities of southern New England, New York, and England.

Registry of Maine Toolmaker Listings

Key to the Registry Listings

- All toolmakers with no asterisk (* or **) were found in the *Directory of American Toolmakers* (DATM) (Nelson 1999). Information on these toolmakers comes from the DATM text, Pollak's (2001) *Guide to American Planemakers*, museum donors, visitors, and other information sources.
- All (*) names were found in Yeaton (2000) but were not located in the DATM.
- Toolmakers located by the museum or other sources are denoted by ** and include the information source or donor's identity.
    - Many references are made to the Bob Jones collection, formerly located in New Mexico, which was dispersed at a Martin Donnelly auction in 2006. The current locations of most of his planes are now unknown.
    - A number of names on planes found in Maine and listed in Pollak (2001) or in the Bob Jones collection are included in this registry and may be Maine planemakers. Additional information on these entries would be greatly appreciated.
    - Toolmakers listed in the *Maine Business Directories* or *Maine Registries* are so noted. All *Maine Business Directories* located by museum staff are in the bibliography. For the sake of brevity, in the listings they are mentioned with the date only instead of including the author.
    - Raymond and Michael Strout of Bar Harbor have made numerous specimens of old advertisements and ephemera available for inclusion in the registry.
    - The gunsmiths of Maine listed in Demeritt's (1973) *Maine Made Guns and their Makers* have been added to the 5th edition.
    - When the Davistown Museum has one or more tools made by a particular toolmaker or manufacturer, the identification number(s) [ID#] and the collection name(s) are listed:
        - MI: Historic Maritime I (1607-1676): The First Colonial Dominion
        - MII: Historic Maritime II (1720-1800): The Second Colonial Dominion & the Early Republic
        - MIII: Historic Maritime III (1800-1840): Boomtown Years & the Dawn of the Industrial Revolution
        - MIV: Historic Maritime IV (1840-1865): The Early Industrial Revolution
        - IR: The Industrial Revolution (1865f.)
- Work dates that are preceded or followed by a dash (-) indicate that they were probably working before or after those dates, but those were the most solid dates attributed to them. If a ca. (circa) precedes a date, it means that this is an estimated

approximate date; the earlier the date the less accurate. For instance, in the case of a particular listing with a work date "ca. 1790", the actual time the toolmaker was working may have been 10 years or more to either side of that date. If the birth or death dates are known, they are located to the right of the work dates or within the text.

- See Appendix G for a preliminary listing of maritime peninsula (Nova Scotia and New Brunswick) edge toolmakers.
- As new information becomes available, biographical and other relevant data will be posted on individual Maine toolmaker web pages with an online link to it from the Registry listing. If you have a biographical sketch or any other important information about any Maine toolmaker, please contact the Museum or mail us your biographical/bibliographic data, and we will post it under the appropriate name. Needless to say, new information on any Maine toolmakers not listed in DATM, Yeaton, or Pollak would be greatly appreciated.

Name	Town	Work Dates	Birth/Death Dates

Abbott, E. G.* Kingfield, Knox -1874-1891
Tools Made: Axes
Remarks: He worked in Kingfield from 1874-1882 and Knox from 1882-1891 (Yeaton 2000).

Abbott, James M.** Mechanic Falls 1891-1898
Tools Made: Gunsmith
Remarks: (Demeritt 1973, 163).

Abbott, Ebenezer G.** Montville -1850-1860-
Tools Made: Blacksmith
Remarks: He is listed in the 1850 and 1860 census. He was born in Northport, Maine.

Abbott, Joel** Montville -1860-
Tools Made: Blacksmith
Remarks: He is listed in the 1860 census. He was born in Vermont.

Abbott, Sewall L. Deering -1871-
Tools Made: Bits
Remarks: Abbott had a patent marked: **PAT' D. JULY 25. 71** for a countersink, but he may not have actually made the tool himself.

Ackerman & Johns* Bangor
Tools Made: Axes

Adams, Dummer J.** Kittery -1877-1900-
Tools Made: Countersinks and Gunsmith
Remarks: The Davistown Museum has in its IR collection (ID# 32405T1) a countersink marked **PATENTED | JAN 23, 1877. | D.J. ADAMS | KITTERY, ME.** with the word "patented" upside down. The Directory of American Machinery and Tool Patents website lists this as patent number 186,513 for an improvement for countersinks and also shows the patent diagram online at: http://www.datamp.org/displayPatent.php?number=186513&type=UT. It is unknown who manufactured the countersink. This one is also marked **R.L. MARKS**, who might have been the owner. Demeritt (1973, 163) notes Adams made at least one surviving double shotgun.

Adams, J. C.* Waterboro
Tools Made: Axes

Adams, J. G.   Waterford   -1855-1856-
Tools Made: Axes and Edge Tools
Remarks: He is listed in the 1855 and 1856 *Maine Business Directories*.

Adams, R. F.   Lincoln   -1855-1856-
Tools Made: Edge Tools
Remarks: He is listed in the 1855 and 1856 *Maine Business Directories*.

Albee, Robert**   Bucksport   -1855-
Tools Made: Wheelwright
Remarks: He is listed in the 1855 *Maine Registry and Business Directory*.

Alden, H. & Co. **   Camden   -1855-
Tools Made: Pump and Block Makers
Remarks: They are listed in the 1855 *Maine Registry and Business Directory*.

Alden, John**   Augusta   1830-
Tools Made: Coopers' tools
Remarks: John Alden, one of the original Plimouth Pilgrims, was a cooper, and, in the 1830's, he and John Howland were sent up to Kennebec to set up and manage a trading post there. This information courtesy of Ruth DeWilde Major.

Aldrich & Waterhouse   Gardiner   -1867-1871-
Tools Made: Edge Tools
See: Waterhouse, W. H.
Remarks: They are listed in the 1869 *Maine Business Directory*.

Aldrich, Ezra   Calais   -1865-1866-
Tools Made: Adzes, Axes, Chisels, and Edge Tools

Aldrich, Ezra**   Bath   -1879-1882-
Tools Made: Axes, Adzes, etc.
Remarks: He is listed on Commercial St. in the 1879 and Water St. in the 1881-2 *Maine Business Directory*. Possibly this is the same Ezra Aldrich as the earlier one in Calais.

Allard, Isaac   Belfast   -1868-1884   (b.1819; d.1884)
Tools Made: Screwdrivers
Remarks: A watchmaker, machinist, and jewelry dealer, Allard was also the inventor of several spiral screwdrivers for which he received patents (80583 on 4 Aug. 1868 and 157087 on 24 Nov. 1874). These tools were known to be made by F. A. Howard (& Son)

after the 1880's; however, their names do not appear on all such tools. Allard, or someone else, may have made them as well. Screwdrivers bearing Allard's patent were still being sold in 1898. This information is courtesy of C. D. Fales. The Davistown Museum has one screwdriver with an Allard patent in the IR collection (ID# 111001T25). Also see Franklin Augustus Howard and Clifford Fales' (1992) *Gristmill* article on Isaac Allard spiral screwdrivers.

Allen, A.**   East Livermore   -1882-
   Tools Made: Edge Tools
   Remarks: He is listed in the 1882 *Maine Business Directory*.

Allen, Cyrus K.   Windham   -1871-1879-
   Tools Made: Farm Tools and Plows

Allen, I. F.**   Liberty   -1870-
   Tools Made: Machinist
   Remarks: He is listed in the 1870 census as living at the home of blacksmith Richard Gilman.

American Axe & Tool Co.   Oakland   -1889-1921
   Tools Made: Axes, Edge Tools, Farm Tools, and Scythes
   Remarks: This was a huge conglomerate formed by 14 American ax manufacturing companies. They eventually acquired some of Maine's ax-makers. Hubbard & Blake of Oakland was one of the original 14 founding companies as was Hubbard & Co. of Pittsburgh, PA. There was no prior connection between the two "Hubbard" companies. The A. A. & T. parent company was originally headquartered in three places: Troy, NY, Boston, MA, and Philadelphia, PA. Those offices were discontinued on Oct. 15, 1890 with all communications directed to New York while their Glassport, PA, manufacturing campus was being constructed. The Oakland facilities became American Axe & Tool Co. plant number 16. In some cases the original companies provided more than one manufacturing plant to the overall effort resulting in more functioning plants than original companies. The A. A. & T. Co. was actually an ax-making trust. Their most fervent competition initially was the Dunn Edge Tool Co. of Oakland, Maine. In 1921, Kelley Axe Mfg. Co. of Charleston, West Virginia (also known as the Kelley Axe & Tool Co.) bought out the A. A. & T. Co. along with all their assets, including facilities, patents, and brands. This information is courtesy of Tom Lamond.

Ames, Epthah J.*   Waterboro
   Tools Made: Axes
   Remarks: Possibly this is the same man as Ames, Jepthah and/or Ames, J.

Ames, George**   Portland   1823-1834
Tools Made: Gunsmith
Remarks: (Demeritt 1973, 30-1, 163).

Ames, J. F.*   Richmond   -1855-1902

Tools Made: Axes and Edge Tools
Remarks: He is listed in the 1855 *Maine Business Directory* as an edge toolmaker. Yeates (2000) reports him as an ax-maker from 1874 to 1902. Perhaps this is the same J. F. in the partnership of J. F. & D. C. Ames. The Davistown Museum has an Ames peen adz marked **J. F. AMES** in the MIV collection (ID# 020807T1). This tool is part of the Art of the Edge Tool Show.

Ames, J. F. & D. C.   Richmond   -1867-1881-
Tools Made: Edge Tools, Axes, Ice Tools, and Shipsmith
Remarks: They are listed in the 1867, 1869, 1879, and 1881 *Maine Business Directories*.

Ames, J.*   Waterville   1826-
Tools Made: Axes

Ames, Jepthah   Waterville
Tools Made: Axes

Ames, M.**
Tools Made: Planes
Remarks: Pollak (2001, 20) lists a 15" gutter plane marked **M.AMES** found on the Maine coast.

Andrews, Edgar D.   Stow   -1855-1882-
Tools Made: Edge Tools
Remarks: DATM (Nelson 1999) indicates Andrews was also working in North Chatham, NH, at the same time. He is listed in the 1881 and 1882 *Maine Business Directories* in both places.

Andrews, W. & Co.   Buckfield Village   -1849-
Tools Made: Hoes

Archer, Joel W.   Lincoln   -1855-1874-
Tools Made: Axes and Edge Tools
Remarks: He is listed in the 1855 and 1856 *Maine Business Directories*.

Arnold**   -1790-

Tools Made: Planes

Remarks: Pollak (2001, 23) lists a 10" beech molder with flat chamfers and a relieved wedge found in Maine. The plane has the appearance of planes made by Jo. Fuller or Arnold & Field, both of Providence, RI. (Note: there is no signature on the plane.)

Atkins & Simmons**   Rockland   -1869-

Tools Made: Shipsmith

Remarks: They are listed in the 1869 *Maine Business Directory*.

Atkins, J. W.*   Rockland   1883-1892

Tools Made: Axes

Atkins, W. J.**   Rockland   -1881-1882-

Tools Made: Shipsmith

Remarks: He is listed in the *Maine Business Directory* in 1881 on Ocean St. and in 1882 at 16 North Main St.

Atkinson, Joseph*   Buxton   1829-1832

Tools Made: Axes

Remarks: This maker is listed as Alkinson in DATM (Nelson 1999), probably a misspelling.

Augusta Foundry**

Remarks: No further information or working dates are available about this listing.

Auster & Davis   Addison   -1856-

Tools Made: Edge Tools

Remarks: This company is listed in the 1856 *Maine Business Directory*.

Austin, Moses**   Addison   -1855-

Tools Made: Edge Tools

Remarks: He is listed in the 1855 *Maine Business Directory*.

Austin, R. J.   Dixfield

Tools Made: Lumber Measuring Tools and Log Calipers

Avery & McLaurin**   Portland   -1882-

Tools Made: Shipsmith

Remarks: They are listed in the 1882 *Maine Business Directory* at 17 Union St.

Avery, George**   Portland   -1881-
Tools Made: Shipsmith
Remarks: He is listed in the 1881 *Maine Business Directory* at 4 Union St.

Avery, George**   Prospect   -1872-
Tools Made: Blacksmith
Remarks: Along with George Wescott, Avery owned and ran Bowdoin Point Blacksmith. The hill upon which the blacksmith shop sat is still referred to as Blacksmith Shop Hill. They were connected with the quarries (Ellis 1980, 251-2).

Avery, Samuel**   Portland   -1869-
Tools Made: Shipsmith
Remarks: He is listed in the 1869 *Maine Business Directory*.

Avery, Samuel**   Charlemont
Tools Made: Axes
Remarks: Avery is listed in Appendix 2 of Klenman's (1990, 98) *Axe Makers of North America* with a notation that Avery was a working blacksmith. Klenman does not give any further description of Avery in his text.

Avery, Stephen   Amherst, North Anson, Anson   -1849-1879-
Tools Made: Edge Tools, Locksmith, and Gunsmith
Remarks: Avery later moved to North Anson, ME. He is listed in North Anson in the 1855 *Maine Business Directory*. The 1856 *Maine Business Directory* lists Stephen Avey of Anson, probably a misprint of his name. He is also listed by Demerrit (1973, 163) as working until 1879.

Ayer, William**   Liberty   -1879-
Tools Made: Axes
Remarks: He is listed in the 1879 *Maine Business Directory*.

Ayers*   Dresden   1869-
Tools Made: Axes

Ayman, William**   Calais   -1855-
Tools Made: Pump and Block Makers
Remarks: He is listed in the 1855 *Maine Registry and Business Directory*.

Babcock, M.**   ?
Tools Made: Drawshaves
Remarks: A mast shave (?) 19" long with a 12 3/8" long cutting blade was found in a

coastal Maine workshop, and the maker is not listed in DATM (Nelson 1999). This shave is in the Davistown Museum collection and is characterized by a heavy cutting blade, 2" in depth; welded steel construction with evidence of heavy filing, appearance 1840 - 1860. It is uncertain if this is a heavy duty coopers' shave, or, as is more likely, a mast shave. If not of Maine origin, it is most certainly a New England-made edge tool.

Bachelder, E. S.** Montville -1869-
Tools Made: Edge Tools
Remarks: He is listed in the 1869 *Maine Business Directory*. See Edward S. Batchelder.

Bacheldor, L.** Sebago
Tools Made: Planes
Remarks: An 8 ½" smooth plane with marks **J. BRADFORD. PORTLAND** and **L. Bacheldor. Sebago** is in the Bob Jones collection (Pollak 2001, 29). Also see Joseph Bradford of Portland and I. Batchelder, who made planes in Sebago. Possibly this is the same person with the name misspelled.

Bailey & Evans** Portland 1850
Tools Made: Gunsmith
Remarks: Demeritt (1973, 163) indicates this is Freeman Evans of 233 Fore St.

Bailey, Edward H. (See Darling & Bailey)

Bailey, G. E. Cambridge -1879-
Tools Made: Farm Tools and Plows

Bailey, Gilbert L.** Portland -1850-1904
Tools Made: Gunsmith
Remarks: Demeritt (1973, 163).

Bailey, Jesse Dresden -1879-
Tools Made: Farm Tools, including Hay Presses

Bailey, Lebbeus** Portland, North Yarmouth -1820-1850-
Tools Made: Gunsmith and Clockmaker
Remarks: (Demeritt 1973, 31, 33-4, 163).

Bailey, R. G. North Bridgewater -1849-
Tools Made: Edge Tools

Bailey, Richard   Bridgton   1816-1856
Tools Made: Edge Tools
Remarks: He is listed in the 1855 and 1856 *Maine Business Directories* as an edge toolmaker.

Bailey, S. Everitt   Cambridge   -1879-
Tools Made: Farm Tools and Plows
Remarks: It is unknown what Everitt's relationship was to G. E. Bailey.

Baker & Kincaid**   Skowhegan   -1867-
Tools Made: Butcher Knives
Remarks: This company is listed in the 1867 *Maine Business Directory*.

Baker, Amon**   Moscow   1880-1881
Tools Made: Driving Calks
Remarks: He is listed by Hoyt (1881).

Baker, C. M. & A.*   Bingham   -1870-1890
Tools Made: Axes, Shaves, and Driving Calks
Remarks: This company is listed in the 1879 *Maine Business Directory*. It is also listed in Hoyt (1875 and 1881). Moore (1940) gives the following description of Main St. in Bingham in 1870:

> Cyrus N. Baker… seemed to have a mania for sidehill structures. He built the machine shop that once stood on the high bank of the Austin Stream a short distance above the old bridge, where he had trip-hammers for making the once-famous Baker driving (shoe) calks, and where his family lived in an apartment above the din of the hammers. He later built the side-hill house in which Steve Clark is now living; and following that he built an office building for the Baker Mfg. Co., on the sidehill near the road leading up to his house. (Moore 1940)

Baker, Cyrus**   Moscow   1880-1881
Tools Made: Driving Calks
Remarks: He is listed by Hoyt (1881).

Baker, D. I.**   South Weston, Weston   -1879-1882-
Tools Made: Axes
Remarks: He is listed in the 1879, 1881, and 1882 *Maine Business Directories*.

Baker, Daniel**   Liberty   -1880-
Tools Made: Blacksmith
Remarks: He is listed in the 1880 census.

Baker, H.   Portland   -1869-
   Tools Made: Saw Filing Machines and Saw Tools

Baker, R.**   York   1997-
   Tools Made: Planes (and Reproductions)
   Remarks: Bob Baker worked in Michigan from 1979 - 1997 and now has moved to Maine. He makes planes for violin makers in New York City and San Francisco. He has also done reproductions of planes, including a Sandusky center-wheeled plow owned by Lee Valley Tools and illustrated in their 2002 calendar (photo) and an Israel White of Philadelphia plow plane illustrated in a Don Rosebrook (2003) publication on plow planes.

Photo courtesy of Lee Valley Tools Ltd.

Baker, Sanford J.**   Oakland   1898
   Tools Made: Scythes
   Remarks: Tom Lamond (personal communications) notes that Sanford J. Baker of Oakland has Pat. No. 615,518, Dec. 6, 1898 for a scythe. It is shared with John King, also of Oakland.

Baker, William G.**   Steuben   -1870-
   Tools Made: Gunsmith
   Remarks: (Demeritt 1973, 163).

Bangor Edge Tool Co.*   Bangor   -1873-1946
   Tools Made: Axes and Knives
   Remarks: This company is listed at 57 or 59 Exchange St. in the 1879, 1881, and 1882 *Maine Business Directory*. Several pieces of ephemera have recently been discovered by Bar Harbor collector Michael Strout and are shown below. The envelope clearly denotes this company as a vendor of the Peavey cant hook and, it is assumed, other tools. The letterhead also has in the name C. A. Peavey in the upper left hand corner above the cant hook illustration, linking the Bangor Edge Tool Co. to the labyrinth of the Peavey clan of toolmakers. The receipt and ax label also include a reference to Peavey Manufacturing

Co. A later (1941) catalog issued by the Peavey Mfg. Co. is reproduced in the Davistown Museum Maine toolmaker information files.

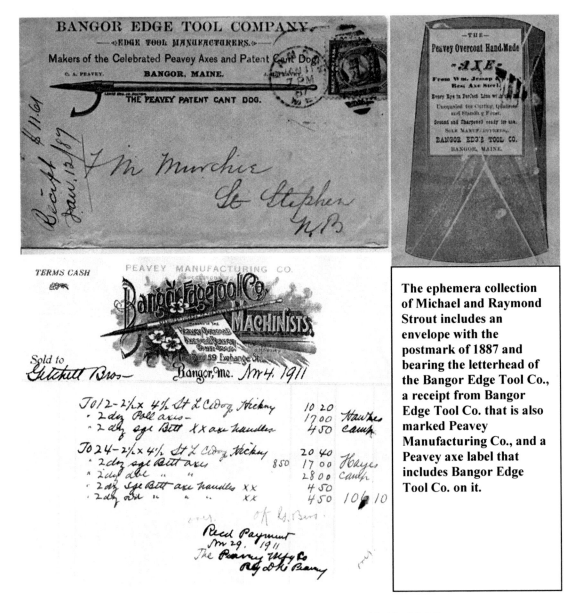

The ephemera collection of Michael and Raymond Strout includes an envelope with the postmark of 1887 and bearing the letterhead of the Bangor Edge Tool Co., a receipt from Bangor Edge Tool Co. that is also marked Peavey Manufacturing Co., and a Peavey axe label that includes Bangor Edge Tool Co. on it.

Bangor Foundry & Machine Co.   Bangor   -1880-1885-
   Tools Made: Farm Tools and Plows
   Remarks: They also manufactured cultivator teeth.

Banks, Edward Prince   Portland   -1834-1870
   Tools Made: Scientific Instruments, including Watches and Mathematical and Nautical Instruments
   Remarks: DATM (Nelson 1999) indicates that Banks was a partner with William H.

H. Hatch in the jewelry business from 1837 to 1857, during which time he may have made instruments, the maker's sign for which is unknown.

Barker & Stinchfield** Brunswick -1825-
Tools Made: Blacksmith

Barker, Abijah** Canaan -1867-
Tools Made: Edge Tools
Remarks: He is listed in the 1867 *Maine Business Directory*.

Barker, Daniel J. Weston -1867-1879-
Tools Made: Axes
Remarks: He is listed in the 1867 and 1869 *Maine Business Directories*. Possibly he could have later partnered in Kierstead & Barker of nearby Danforth.

Barclay, David** Calais -1855-
Tools Made: Wheelwright
Remarks: He is listed in the 1855 *Maine Registry and Business Directory*. It also has the word "milltown" in parentheses next to his name.

Barclay, Samuel** Calais -1855-
Tools Made: Wheelwright
Remarks: He is listed in the 1855 *Maine Registry and Business Directory*.

Barney, H. W.** Waterville 1856-1860-
Tools Made: Gunsmith
Remarks: (Demeritt 1973, 163).

Barnes & Smith** Brewer -1869-
Tools Made: Shipsmith
Remarks: He is listed in the 1869 *Maine Business Directory*.

Barr, H. G.** Skowhegan -1870-
Tools Made: Gunsmith
Remarks: (Demeritt 1973, 163).

Barrett, G. Buckfield -1879-1885-
Tools Made: Farm Tools and Rakes, including Drag Rakes

Barrows, C. & Son Canton -1871-
Tools Made: Horse Rakes

Barrows, A.   Canton   -1879-1885-
Tools Made: Farm Tools
Remarks: His relationship to Barrows & Son is unknown.

Barrows, L. M.**   Vassalboro
Tools Made: Planes
Remarks: He made a 9" long compassed smoother, marked with his name and
VASS.ME. The town of Vassalboro, Maine, located on the east bank of the Kennebec
River, 14 miles northeast of Augusta, had a population of 3,000 in 1850 (Pollak 2001,
36). The Bob Jones collection has a 9" iron coffin plane with this mark.

Bartlett, J. W.   Elliot
Tools Made: Branding Irons
Remarks: DATM (Nelson 1999) indicates, "It is not clear if Bartlett was it's maker or
the name it branded."

Barstow, Calvin**   Brunswick   -1790-
Tools Made: Blacksmith

Barton, Flint**   Oakland   1773-1833
Tools Made: Axes
Remarks: Barton was one of the first ax-makers working on the Emerson Stream in
Oakland, Maine (Klenman 1990). Klenman indicates that he had 13 sons, many of whom
became blacksmiths as well as ax-makers in the Oakland area. No specific names or
manufacturing company records about the activities of these offspring are known to have
survived.

Bassell, T. H.   Wellington   -1870-1871-
Tools Made: Axes

Batchelder, Edward S.   Montville   -1869-1871-
Tools Made: Blacksmith and Edge Tools
Remarks: This name is also recorded as Bachelder. He is listed in the 1850 - 1870
censuses and was born in Liberty.

Batchelder, I.   Sebago   -1790-1800-
Tools Made: Wood Planes
Remarks: DATM (Nelson 1999) indicates that the maker's signature was **I.**
BATCHELDER | SEBAGO. Pollak (2001, 40) states that I. Batchelder was listed in
Sebago ca. 1800. Also see L. Bacheldor.

Batchelder, Joseph C.   Lisbon   -1819-1828-
Tools Made: Edge Tools and Scythes
Remarks: It is unclear (Nelson 1999) if a hay knife with the reported marking
BATCHELDER may also have been from this maker.

Bates, Erastus W.**   Waterville   -1863-
Tools Made: Handsaws
Remarks: He had patent 37999 on 3/24/1863 for a handsaw. It was assigned to John
Ellis of N. Bridgewater, MA. The patent claims an improvement to the method of
stretching the blade in a buck saw. See Ellis Saw Co. Source: Graham Stubbs.

Bath Iron Foundry   Bath   -1869-1871-
Tools Made: Farm Tools and Plows, including Cultivators and Harrows.

Baxter, A. P.**   Waterville
Tools Made: Gunsmith
Remarks: (Demeritt 1973, 105, 163).

Bead & Morrill**   Bangor
Tools Made: Planes
Remarks: Morrill was a prominent Bangor planemaker. It is not known who "Bead"
was (Pollak 2001).

Beam, Eli B.   Brownfield   -1885-
Tools Made: Farm Tools

Bean & Day**   Biddeford   1874-1880
Tools Made: Gunsmith
Remarks: Demeritt (1973, 163) notes that this is a partnership of Samuel E. Bean and
Benjamin F. Day.

Bean, B. B.*   Rockland   -1869-
Tools Made: Shipsmith
Remarks: He is listed in the 1869 *Maine Business Directory*.

Bean, J.**   Old Town
Tools Made: Gunsmith
Remarks: (Demeritt 1973, 163).

Bean, Jonathan   Montville   -1855-1860-
Tools Made: Farm Tools and Machinist

Remarks: He is listed in the 1855 *Maine Registry and Business Directory* and in the 1860 census as a machinist. He was born in New Hampshire.

Bean, Samuel E.** Biddeford 1866-1880-
Tools Made: Gunsmith
Remarks: See Bean & Day (Demeritt 1973, 163).

Beath, W. H.** Kenduskeag -1879-
Tools Made: Axes and Coopers' Tools
Remarks: He is listed in the 1879 *Maine Business Directory*.

Beek & Sons* Orono -1874-1877
Tools Made: Axes

Belfast Foundry** Belfast -1880-1900-
Tools Made: Tools and Machinery
Remarks: The Belfast Foundry originally did custom foundry work, including ships' fittings. They also made quarry tools, and were known to have sold stone dressing tools to Halls Quarry in Mount Desert. This information was provided by Doug Brown of Belfast, who also has a house jack made by the Belfast Foundry. His grandfather worked there. The foundry made a proof press for a local printer.

Bellmore, C. S.** Skowhegan
Tools Made: Gunsmith
Remarks: Demeritt (1973, 163) notes he was part of Phillips & Bellmore.

Benjamin & Co. Winthrop -1855-
Tools Made: Farm Tools

Benjamin, James Newport -1862-
Tools Made: Axes
Remarks: Benjamin was also listed as a horseshoer.

Benjamin, Samuel** Winthrop -1790-
Tools Made: Planes
Remarks: Pollak (2001, 45) states that he was apprenticed to Joseph Metcalf, a Winthrop, ME, planemaker.

Bennett's Mills Norway -1864-
Tools Made: Farm Tools and Plows
Remarks: George Evans bought the company in 1864, where he made the circular plane that he patented on 28 Jan. 1862. DATM (Nelson 1999) indicates that this company

may have made planes for Evans before he bought the company, and it is not known whether or not Evans continued to make farm tools and plows once he bought the company. For more information on G. F. Evans, see the entry under that name. The Davistown Museum has one specimen of the Evans circular plane on exhibit in the IR collection (ID# TJE1001).

Bennett, Epenetus**   Waterville   1838
Tools Made: Gunsmith
Remarks: (Demeritt 1973, 34-6, 163).

Bennett, F. P. & Co.**   Montville
Tools Made: Blacksmith and Machinist
Remarks: "Frank [Bennett] was an inventor who developed an improved model of the Cram Wheel that had a more efficient turbine action, and was designed to allow for quicker repairs when the iron fins broke off. ...Several of his patents were offered for general sale, such as the Liberty Tongue and Groove Stave Machine and Bennett's Improved Stave Chamfer and Crozing Machine, and a hand water pump" (Donahue 1996). The illustration of the stave chamfer and crozing machine is from Donahue (1996). An illustration of the tongue and groove stave machine is reproduced for the Liberty Machine Co. listing.

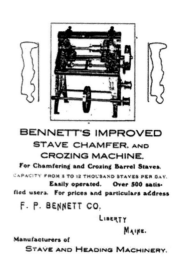

BENNETT'S IMPROVED STAVE CHAMFER, AND CROZING MACHINE.
For Chamfering and Crozing Barrel Staves.
CAPACITY FROM 8 TO 12 THOUSAND STAVES PER DAY.
Easily operated.   Over 500 satisfied users. For prices and particulars address
F. P. BENNETT CO,
LIBERTY
MAINE.
Manufacturers of
STAVE AND HEADING MACHINERY.

Berry & Waterhouse*   Gardiner
Tools Made: Axes

Berry, Eldbridge**   Gardiner   -1867-
Tools Made: Edge Tools
Remarks: He is listed in the 1867 *Maine Business Directory*.

Best, Henry**   Robbinston -1855-
Tools Made: Pump and Block Makers
Remarks: He is listed in the 1855 *Maine Registry and Business Directory*.

Bicknell Manufacturing Company**   Rockland
Tools Made: Quarrying
Remarks: Mr. Putnam Bicknell, the firm's second manager, purchased control of the Livingston Manufacturing Company and opened a branch office and plant in Elberton, Georgia. He also changed the name to Bicknell Manufacturing. The company now goes by the name Bicknell Supply Co. and has closed the Rockland branch. For more

information, see the Davistown Museum's Bicknell information file. The Davistown Museum has a Bicknell stone chisel (ID# 12801T11) in the IR collection.

Billings**   Augusta   -1845(?)-1870(?)

Tools Made: Drawknives, Cleavers, and Slicks
Remarks: The Davistown Museum has a clapboard slick (ID# 52403T3) marked **BILLINGS | AUGUSTA** in the MIV collection (photo) and a cleaver (ID# 61601T1) in the MIII collection.

Billings**   China
Tools Made: Chisels
Remarks: The Davistown Museum has a chisel (ID# 81602T17) marked **BILLINGS. | CAST STEEL | CHINA** and **CAST STEEL | WARRANTED** in the MIII collection (photo).

Billings Axe Co.**   Clinton
Tools Made: Axes
Remarks: This company was founded by J. P. Billings of Clinton. George P. Billings was born in 1865 and eventually took over his father's business.

Billings & Fogg   North Monmouth   -1859-
Tools Made: Hoes and Shovels

Billings & Spear   North Monmouth   -1849-1856-
Tools Made: Hoes and Shovels
Remarks: DATM (Nelson 1999) indicates that they were succeeded by Billings & Fogg sometime before 1859 and were also listed as Spear & Billings.

Billings, George H.   North Monmouth   -1869-1871-
Tools Made: Hoes and Shovels

Billings, G. P.*   Clinton   -1874-1909
Tools Made: Axes
Remarks: Connor Noel has informed us that John P. Billings of Clinton had a son named George P. Billings, born in 1865. This is possibly the G. P. Billings listed by Yeaton (2000).

Billings, J.*   Saco   -1874-
Tools Made: Axes
Remarks: Possibly he could be a relative of John P. Billings of Clinton.

Billings, John P.**   Clinton   1837-1881-   b. 1837
    Tools Made: Axes and Edge Tools
    Remarks: DATM (Nelson 1999) notes his maker's mark as **J.P. BILLINGS |
CLINTON. ME** with both lines of text curved to form an oval. DATM also notes a John
Billings worked in Saco, Maine circa 1825, as well as in Hallowell in 1841. Connor Noel
(a sixth generation descendent of John P.
Billings) has informed us that John P. was
born in 1837, and that he opened the ax
factory in Clinton (dates unknown). This
means that the date of 1825 given in
DATM is not correct or that it refers to yet
another of the Billings clan. He is listed as

a Clinton ax-maker in the 1867, 1879, and 1881 *Maine Business Directories*. The broad
ax shown (with an enlargement of the mark) belongs to Connor Noel. The Davistown
Museum has a Billings ax (ID# 121600T1) listed in the IR collection and a hewing ax
listed in the MIV collection (ID# 42604T5). On 10/28/05, the Hulls Cove Tool Barn
discovered a **J.P. BILLINGS | CLINTON MAINE** lipped adz (MIV collection, ID#
121906T2) in a Falmouth foreside barn, the first one noted in the 38 years of tool picking
by the Jonesport Wood Co. It is included in the Art of the Edge Tools show.

Bisbee, Asa**   North Yarmouth   1810   (d.1865)
    Tools Made: Gunsmith and Blacksmith
    Remarks: (Demeritt 1973, 163).

Bishop, George**   Morrill's Corner, Westbrook   1870-1907
    Tools Made: Gunsmith
    Remarks: (Demeritt 1973, 105, 163).

Blackwell, O.**   Skowhegan   1873
    Tools Made: Gunsmith
    Remarks: (Demeritt 1973, 163).

Blake, R.**   Center Montville   1877
    Tools Made: Gunsmith
    Remarks: (Demeritt 1973, 163).

Blake, Robie**   Cornish   1894
    Tools Made: Nail Extractor
    Remarks: He has Pat. No. 526. 678, Oct. 2, 1894 for a nail-extractor (Tom Lamond
personal communication).

Blake, Wm. P.** Waterville 1862-1889
Tools Made: Axes, Scythes, and Hooks
Remarks: See the Hubbard & Blake Mfg. Co.

Bliss, T. Buckston, Bucksport -1807-1820-
Tools Made: Wood Planes
Remarks: The definite listing of Bliss as a toolmaker in the first edition of the DATM (Kijowski 1990) has been changed to a tentative listing in the 1999 edition (Nelson 1999). The 1999 edition also notes that Bliss may be one of two cabinet makers in Newport, RI. Planes were simply marked **T. BLISS**. Pollak (2001, 52) notes the mark **T. BLISS. / BUCKSTON** (in 1807 Buckston's name was changed to Bucksport.) The Bob Jones collection has a 9 ¼" round plane with the Buckston mark.

Blodett, S. A.** Belfast -1881-
Tools Made: Shipsmith
Remarks: He is listed in the 1881 *Maine Business Directory* on Main St.

Bodwell, Joseph** Wayne, Oakland 1879
Tools Made: Axes
Remarks: See the North Wayne Tool Co.

Bolton & Son Portland
Tools Made: Axes
Remarks: It is unknown if they had any relationship to Libby & Bolton, Portland's prolific maker of edge tools.

Bolton, Elbridge G. Portland -1855-1856-
Tools Made: Adzes, Drawknives, and Edge Tools
Remarks: DATM (Nelson 1999) indicates his maker's marks as **E.G. BOLTON PORTLAND** and **E.G. BOLTON | FLOYD | PORTLAND. ME** with a notation that Floyd was otherwise reported as a Portland ax-maker. He is listed in the 1855 and 1856 *Maine Business Directories*. Also see Libby & Bolton. Note: Clarence Blanchard sold both a signed Bolton hewing ax and a Bolton gutter adz at his auction on March 10, 2002.

Bolton, Thomas Portland -1849-
Tools Made: Drawknives and Edge Tools
Remarks: His maker's mark is noted as **T. BOLTON**.

Booker, W. H.** Gardiner 1892
Tools Made: Gunsmith
Remarks: (Demeritt 1973, 163).

Boothby, David S.**   Livermore, East Wilton   -1870-1890-

Tools Made: Knives and Gunsmith
Remarks: D. S. Boothby is listed as making knives in
the 1882 *Maine Business Directory*. Demeritt (1973, 163)
lists him as a gunsmith in Livermore starting in the 1870s and then moving to East
Wilton in the 1890s. Steve Beauregard of Hinckley, Maine has a 16 ½" knife marked **D.S.
BOOTHBY EAST WILTON** (photograph courtesy of Rick Floyd).

Boothby, Edward K.**   Portland   1852-1890-   (d. 1899)
Tools Made: Gunsmith
Remarks: (Demeritt 1973, 92, 163).

Boothby, Noah   Waterville   -1881-1882-

Tools Made: Axes and Edge Tools
Remarks: He is listed in the 1881 and 1882 *Maine Business
Directories* with a location of Main St. He used the mark **N.
BOOTHBY | WATERVILLE**. The 2" socket chisel in the
photograph is owned by Rick Floyd.

Boothy, Brice   Limington   -1855-1856-
Tools Made: Edge Tools
Remarks: He is listed in the 1855 *Maine Business Directory* with a first name of
"Price" and in the 1856 *Maine Business Directory* as "Brice."

Bore, R & H, Co.**   Waterville
Tools Made: Axes

Boulet**   Sebago Lakes   -1900-1904-
Tools Made: Gauges
Remarks: The Davistown Museum has a measuring
gauge (ID# 102503T1) marked **BOULET'S FINE TOOL WORKS SEBAGO LAKES
MAINE | PAT OCT 2 1900 SEPT 10 1910 FEB 25 04** in the IR collection (photo).

Boulton, N. L. V.   Waterville   -1849-
Tools Made: Hammers

Bowker, James, Sr.**   Paris   1796-1820-
Tools Made: Blacksmith
Remarks: He purchased land in Paris, ME, in 1796 from Isaac Jackson. He was a
descendant of James Bowker, a Swede in early Massachusetts and was also a
Revolutionary Soldier.

Boynton, D. P.   Monmouth   circa 1878-
Tools Made: Wood Bench Vises
Remarks: According to DATM (Nelson 1999), it is unclear if Boynton patented a wood bench vise in 1878, made it, or perhaps both.

Bradford, Jesse   Turner   -1869-
Tools Made: Farm Tools and Plows

Bradford, Joseph   Portland   circa 1837-1884   (b. June 1806, d. May 21, 1884)
Tools Made: Carpenter Tools, Cooper Tools, Edge Tools, Shipsmith Tools, and Wood Planes
Remarks: He had two marks: **J.BRADFORD | PORTLAND** and **J.BRADFORD | PORTLAND | ME**. According to Pollak (2001, 58), "The 1860 census reported he had one employee and produced $1000 worth of coopers' tools, $200 worth of joiners' tools and $100 in repairs." The Bob Jones collection has four examples of his planes, including an 11 ½" beech-handled match grooving plane. The sketch, courtesy of Bob Jones, is of his shipbuilder's rabbet. Joseph Bradford is listed in the 1869 *Maine Business Directory* at 200 Fore St.

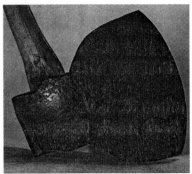

J. BRADFORD
PORTLAND
ME

E H M
SHIPBUILDERS RABET Squared 10½'
Round chamfers
Birch

No marks on heel

IRON w/ Chip Breaker
Both irons - PROV. TOOL CO
CAST STEEL

Bradley, Peter H.**   Portland   -1869-
Tools Made: Adzes
Remarks: Bradley designed and made an improved adz with a two piece blade, which could be pressed over the handle and held in place by a wood or metal plug. No known examples of this adz survive.

Bragg, H.**   Cornville
Tools Made: Axes
Remarks: The Davistown Museum has a broad ax (ID# 062603T1) marked **H.BRAGG | CORNVILLE** in the MIV collection (photo). It is unknown if there was a relationship between this Bragg and the other Braggs listed below.

Bragg, H. A.   Katahdin   -1856-
Tools Made: Edge Tools
Remarks: He is listed in the 1856 *Maine Business Directory* as an edge toolmaker.

Bragg, N. H., & Sons**   Bangor   -1889-
Tools Made: Blacksmith Tools
Remarks: Figure of bill courtesy of
Raymond Strout.

Bragg, William B.   Skowhegan   -1871-
Tools Made: Farm Tools and Shovels

Bragg, Z. O.**   Rockland   -1869-
Tools Made: Shipsmith
Remarks: He is listed in the 1869 *Maine Business Directory*.

Breck & Weymouth   Clinton   -1856-
Tools Made: Edge Tools
Remarks: This company is listed in the 1856 *Maine Business Directory* as an edge
toolmaker.

Breck, William D.*   Clinton   1855-1866
Tools Made: Axes
Remarks: It is unknown if he was the son of William D. Breck of Waterville or if this
William D. Breck associated with Breck & Weymouth.

Breck, William D.   Waterville   -1800-
Tools Made: Axes

Brett, Rufus*   Phillips   -1855-1856-
Tools Made: Axes and Edge Tools
Remarks: He is listed in the 1855 and 1856 *Maine Business Directories*. Don Yeaton
has recently (2003) found an ax signed **Rufus Brett Phillips**.

Brewer
Tools Made: Axes

Bridges, J.**   Ellsworth   -1856-
Tools Made: Shipsmith
Remarks: He is listed in the 1856 *Maine Business Directory*.

Bridges, John R.**   Castine   1855-1880-
Tools Made: Gunsmith and Blacksmith
Remarks: (Demeritt 1973, 163).

Brigdon* Bridgdon 1816-1832
 Tools Made: Axes

Briggs, S. E. Caribou -1899-1900-
 Tools Made: Farm Tools

Broad, E.** Bangor -1855-
 Tools Made: Slicks
 Remarks: He is listed in the 1855 *Maine Business Directory* as an edge toolmaker. A 16" long slick has been found (photo) marked **E. BROAD | BANGOR | WARRANTED**. The close-up of the mark is from a drawshave. DATM (Nelson 1999) lists several Broads who made edge tools working from 1857 to 1901 in New Brunswick, Canada. See Appendix G on Canadian Toolmakers. It is unknown if there is any relationship between the earlier E. Broad working in Bangor in mid-century and the Canadian Broad clan.

Broad, Hollis China -1856-
 Tools Made: Axes

Brock, William D.** Clinton -1855-
 Tools Made: Edge Tools
 Remarks: He is listed in the 1855 *Maine Business Directory* as an edge toolmaker. It is unknown if this is a different person from the William D. Breck of Clinton, who made axes.

Brodigan, L. Biddeford -1869-1871-
 Tools Made: Files

Brooks, C. C., Arms & Tool Co.** Hallowell, Portland, East Wilton -1870-1900-
 Tools Made: Gunsmith and Hay Knives
 Remarks: Demeritt (1973, 155, 163) indicates that Chapin C. Brooks had an 1880 patent for a reversible plow and an 1882 patent for a piece of machinery for a sawmill from when he lived in Lancaster, NH. He received a patent for a hay knife, July 17, 1888, by which time he was living in Hallowell, Maine. His first gun patent was in 1890 for a combination over and under rifle and shotgun. These are marked **C. C. Brooks East Wilton Me.** In 1893, a *Business Directory* lists the C. C. Arms & Tool Co. in Portland. He has several other patents.

Brooks Hdw. Augusta -1899-1900-
 Tools Made: Farm Tools

Brown, A. D.   Augusta   -1855-
Tools Made: Shovels

Brown, Cyrus   Fayette   -1855-1856-
Tools Made: Edge Tools
Remarks: He is listed in the 1855 and 1856 *Maine Business Directories* as an edge toolmaker.

Brown, James W.   Woodstock   -1862-
Tools Made: Edge Tools

Brown, John**   Pemaquid, Woolwich, Bristol   1625-1659   (b. 1604, d. 1659)
Tools Made: Blacksmith and Mason
Remarks: Probably Maine's earliest toolmaker and shipsmith. For more information see James Phipps and the Davistown Museum information file on John Brown.

Brown, John   St. Albans   -1855-1856-
Tools Made: Axes and Edge Tools
Remarks: He is listed in the 1855 *Maine Business Directory*.

Brown, John Hamilton**   Liberty   1855-1860-
Tools Made: Gunsmith
Remarks: Demeritt (1973, 153, 163) notes he moved to Newburyport, MA, where he worked from 1868 to the 1880s as Brown Manufacturing Co.

Brown, Levi S.**   Portland   -1860-
Tools Made: Gunsmith and Gas Fitter
Remarks: Demeritt (1973, 163).

Brown, N. A.*   Richmond   1886-1889
Tools Made: Axes

Brown, Otis**   Portland   1852
Tools Made: Gunsmith
Remarks: Demeritt (1973, 163).

Brownson, Herbert S.**   Portland
Tools Made: Screwdrivers
Remarks: Brownson obtained patent 344160, 6/22/1886; information courtesy of C.D. Fales.

Bryant, Alonzo   Montville   -1879-1880
Tools Made: Farm Tools, including Winnowing Mills

Bryant, E. A.**   Bath   -1882-
Tools Made: Edge Tools
Remarks: He is listed in the 1882 *Maine Business Directory* on Water St.

Bryant, Samuel**   Portland   -1855-
Tools Made: Pump and Block Makers
Remarks: He is listed in the 1855 *Maine Registry and Business Directory.*

Buck, Amos**   Eastport   -1856-
Tools Made: Shipsmith
Remarks: He is listed in the 1856 *Maine Business Directory.*

Budge, James Thomas*   Lee   -1840-1890-
Tools Made: Blacksmith
Remarks: "He spent his early days on the farm, and in early manhood learned the blacksmith's trade. After becoming of age he worked at that business about sixteen years in this town. In 1863 he engaged in trade and continued at that business for fifteen years, when he sold out and again went into blacksmithing with his son, which business he is now following" (Godfrey 1882, 893).

Bunker, William J.**   Thomaston   -1850-1880-
Tools Made: Gunsmith and Blacksmith
Remarks: Demeritt (1973, 163).

Burbank & Allamby**   Bangor   -1855-
Tools Made: Brass Founders and Finishers
Remarks: They are listed in the 1855 *Maine Registry and Business Directory.*

Burgin, Lewis**   Eastport   -1856-
Tools Made: Shipsmith
Remarks: He is listed in the 1856 *Maine Business Directory.*

Burkett, N. H.   Union   -1879-
Tools Made: Farm Tools, including Mowing Machines and Horse Products

Burnham, J. B.**   Calais   -1882-
Tools Made: Edge Tools
Remarks: He is listed in the 1882 *Maine Business Directory.*

Burnham, J. H.**   Blue Hill   -1855-
Tools Made: Wheelwright
Remarks: He is listed in the 1855 *Maine Registry and Business Directory*.

Burns & Buxton**   Portland   -1855-
Tools Made: Pump and Block Makers
Remarks: They are listed in the 1855 *Maine Registry and Business Directory*.

Burrowes, Edward T., Co.   Portland   1878-1928
Tools Made: Levels, Rules, and Wooden Planes
Remarks: DATM (Nelson 1999) notes Burrowes, "was primarily a maker of a sliding wire screen patented in 1878; the tools he marked were special types used to install that screen," including a sliding and folding rule patented 24 Mar. 1891. Burrowes made one plane in particular in large quantities, the screen-making plane, which has frequently

appeared in tool chests and collections purchased by the Liberty Tool Co. over the past thirty-eight years. Burrowes' mark is printed on the side of his screen-making plane in large block letters (**E.T. BURROWES CO., MANUFACTURERS, PORTLAND MAINE**, and other variations). According to Pollak (2001, 69-70), the plane bearing the firm name was part of the installation package for the screen. The Davistown Museum has a 12" level (ID# 72002T2) marked **E. T. BURROWES CO. | PORTLAND, ME.** in the IR collection (photo). Twenty-four examples of a brass tag stating **MADE BY | E.T.BURROWES & CO | PULL | WIRE SCREENS | PORTLAND,ME.** are in the Bob Jones collection. This company also made furniture items, such as pool tables and cedar chests. We have received a report of an old crank phonograph in a cabinet marked "Burrowes of Portland, Maine".

Bussel, Ethan*   Wellington   1885-1895
Tools Made: Axes
Remarks: There was also a T. H. Bussell that worked in Wellington starting around 1885; perhaps this was a brother or another relative, and the name has been misspelled.

Bussell, T. H.   Wellington   -1869-1925
Tools Made: Axes, Chisels, Drawshaves, etc.
Remarks: He is listed in the 1869, 1879, 1881, and 1882 *Maine Business Directories*.

Buswell, Turner H.**   Brighton   -1855-1856-
Tools Made: Edge Tools
Remarks: He is listed in the 1855 and 1856 *Maine Business Directories* as an edge

toolmaker. In 1856 his name was spelled Bussell. It is unknown if this was the T. H. Bussell reported later in Wellington or a relative.

Butler & Haines   Bangor
Tools Made: Saws
Remarks: Reported also as Butler and Haynes, the maker's signature was **BUTLER & HAINES | BANGOR, ME | WARRANTED CAST STEEL** with the name line curved.

Butler, Daniel W.**   Liberty   -1860-
Tools Made: Blacksmith
Remarks: He is listed in the 1860 census.

Butterfield, Andrew**   East Wilton   1836-   (d.1866)
Tools Made: Blacksmith
Remarks: He bought his father Issac's shop from Samuel Pease in 1836 and partnered with Calvin Keyes in 1839. This information is from the notes of W. A. "Chet" Sweat of Farmington, ME.

Butterfield, Issac, Jr.**   East Wilton   1803-1817   (b.1750 d. 1817)
Tools Made: Blacksmith and Quarrying Tools
Remarks: His blacksmith shop was sold to Samuel Pease in 1818. This information is from the notes of W. A. "Chet" Sweat of Farmington, ME.

Buzzard, A.   Bangor
Tools Made: Hammers and Stone-working Tools
Remarks: The maker's mark **A. BUZZARD BANGOR** was found on a slater's hammer.

C. R. (See C. Record)

Cameron, W. F., & Co.   Portland   -1879-1880-
Tools Made: Farm Tools

Cammett, Dudley**   Portland   -1855-
Tools Made: Pump and Block Makers
Remarks: He is listed in the 1855 *Maine Registry and Business Directory*.

Campbell, Ambrose S.   Ellsworth   -1855-1904
Tools Made: Axes, Chisels, and Edge Tools
Remarks: His maker's mark is **A.S. CAMPBELL**. DATM (Nelson 1999) lists A. S. Campbell working from 1874 - 1904 and Ambrose Campbell working in 1856, and it is unclear whether these are two people or one. He is listed in the 1855 and 1856 *Maine Business Directories* as Ambrose S. Campbell under the edge tool manufacturers'

category. He is also listed in the 1881 and 1882 *Maine Business Directories* as A. S. Campbell, maker of axes.

Campbell, B. W.   Livermore   -1869-
   Tools Made: Handles and Scythes, including Scythe Snaths

Campbell, Charles F.**   Hallowell   -1860-1880-
   Tools Made: Gunsmith
   Remarks: Demeritt (1973, 163).

Capen, David**   Eastport   -1856-
   Tools Made: Shipsmith
   Remarks: He is listed in the 1856 *Maine Business Directory.*

Carlson, C.**   -1910-
   Tools Made: Planes
   Remarks: Pollak (2001,74) lists a 5 1/2" lignum smoothing plane dated "1910" found on the Maine coast. (Note: There is no signature on the plane.)

Carlton, Charles C.**   Portland   1823-1834
   Tools Made: Gunsmith
   Remarks: Demeritt (1973, 163).

Carpenter, F. E.   Portland   ca. 1870
   Tools Made: Wooden Planes
   Remarks: The marking is ink-stamped and reads **F.E. CARPENTER | 153 ALLEN AVENUE | PORTLAND, ME.** and may be that of a dealer. The example cited in Pollak (2001, 76) is a 10 ½" beech dado from the Bob Jones collection. The sketch is courtesy of Bob Jones.

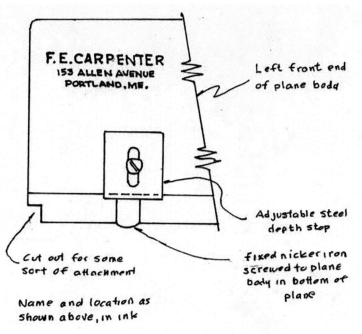

F.E.CARPENTER
153 ALLEN AVENUE
PORTLAND, ME.

Left front end of plane body

Adjustable steel depth stop

fixed nicker iron screwed to plane body in bottom of plane

Cut out for some sort of attachment

Name and location as shown above, in ink

Carr, E. G.**   Dixfield   -1882-
   Tools Made: Edge Tools
   Remarks: He is listed in the 1882 *Maine Business Directory.*

Carr, Hiram   Dexter   -1849-1855

Tools Made: Farm Tools and Handles

Remarks: According to DATM (Nelson 1999), "An 1849 listing is only for scythe snaths; an 1855 one is for agriculture implements. It is not known if the later listing implied he was making anything else."

Cary Bros.   Houlton   -1869-1871-

Tools Made: Farm Tools and Plows

Cary, W. H.**   Houlton   (d. 1859)

Tools Made: Planes

Remarks: Bob Wheeler informed the Davistown Museum that he located a plow plane marked "W. H. Cary" with 1820 stylistic characteristics, in Belfast, Maine. Wheeler indicates that William Holman Cary was born in 1779 in Bridgewater, MA. He married in 1800, relocating to New Salem, MA, and was the father of J. H. Cary. In 1822, he took up land in Houlton, Maine, available to citizens of New Salem under an earlier grant from the Commonwealth of Massachusetts. His son operated a store out of his father's house and went on to own extensive timber interests, in addition to a foundry and carding and grist mills. William was associated with the operation of the various Cary mills. W. H. Cary died in Houlton in 1859, it is not known for certain if he made planes there or not. See the Davistown Museum information file on the Cary family for more information on this family and some of the tools they made.

Cassidy, T. F.**   Bangor   -1873-1896

Tools Made: Cant Dogs, Tackle Blocks, Steering Wheels, Marine Hardware, and Shipsmith

Remarks: This T. F. Cassidy invoice, courtesy of Raymond Strout, was written in 1884. Strout indicates that he was listed as a shipsmith from 1877 to 1921. From 1883(?) to 1921, he is listed as a cant dog manufacturer and holds a cant dog patent dated Dec. 28, 1881. The New England Fire Rescue website also notes, "Jun 04, 1873 - Bangor - ME - near Steamboat Wharf – Cassidy's Cant-dog factory burns." The 1881 and 1882 *Maine Business Directories* lists them as "on Front, opp. May."

Cassidy, T. F., & Son**   Bangor   1897-1922

Tools Made: Cant Dogs, Tackle Blocks, Steering Wheels, and Marine Hardware

Remarks: T. F. Cassidy became T. F. Cassidy & Son in 1897. In 1922, it is only listed as "ship stores & chandlery." The address is given as "32 Front St." and by 1913 as "32 to 42 Front St."

Castner, Wilibaldus**   Broad Bay   1753-1774   (d.1774)
Tools Made: Blacksmith
Remarks: A member of the Castner family provided the following family history. Castner has been anglicized from Kastner or Kestner. Wilibaldus, commonly called Baltas, Baltus, Balthasar, and his wife Augustina, commonly called Justina, came to Broad Bay in 1753 from Konigsbach in Baden Durlach, Germany. Baltas was a blacksmith by trade (signed letter to Bishop Spangenberg, Bethlehem, May 22, 1767). He took up the first lot on the west side of the river in the area of the old Lovell Bridge. After the last Indian War he gave up this lot and with his son, Ludwig, squatted on a vacated lot, the one above the farm now owned by Merle C. Castner. This move was not entirely successful, since when the lot was vacated it reverted to the proprietor. This was Lot No. 11, recently Walter Boggs place, and the Castners were compelled to repurchase it in 1774 of Samuel Waldo's son-in-law, for 13 pounds, 6 shillings, 8 pence. Baltas died in 1774, leaving an estate valued at 110 pounds, 13 shillings, 3 pence. According to family tradition he was buried in the old Lutheran Cemetery on the shore of Merle Castner's farm. There were three children in the Castner family of whom any record has been preserved. Johann Anton, born at Konigsbach, November 29, 1743, moved to North Carolina in 1769 and joined the Moravian congregation in Bethabara. He was of a stormy nature and this led to his exclusion from the church, but he was reconciled to the Brethren again at his death. He married four times and had eleven children. Sophia Salome Castner, a daughter of Baltas, was born April 19, 1734, at Konigsbach. She married Johann Georg Lagenauer who had come to Broad Bay in 1753, and on his death in 1757, she married Friedrich Kuenzel. They migrated to Friedland, North Carolina, in 1770 and remained there until her death March 10, 1816. Ludwig Castner, born at Konigsbach in 1751, remained at Broad Bay and became the progenitor of the Waldoboro Castners. He lived at his father's place and around 1790 built the Old Castner Homestead. Ludwig died January 14, 1822 (Stahl 1956, 258-9).

Caswell, N. N.   Harrison   -1862-
Tools Made: Carpenter Tools and Woodworking Machinery
Remarks: Caswell made machines for producing shingles and staves; it is not known if these were powered machines.

Caswell, H. W.**   Yarmouth   -1867-1879-
Tools Made: Post-hole Augers
Remarks: He is listed in the 1879 *Maine Business Directory*. The Yarmouth Historical

Society has one of his post hole augers in their Henry Caswell Collection, Coll. 140 and provided the museum with a copy of this ad.

Cate, Johnathan**   Limington   -1855-
Tools Made: Pump and Block Makers
Remarks: He is listed in the 1855 *Maine Registry and Business Directory*.

Cates, Nathan A.   Thorndike   -1869-1880-
Tools Made: Farm Tools and Hoes
Remarks: In 1869, Nathan A. Cates made cultivators and horseshoes in Thorndike; it is possible that an N. A. Cates working in Unity from 1879-80 is the same man.

Catland, Thomas R.**   Lewiston   -1880-1890
Tools Made: Gunsmith, Machinist, and Locksmith
Remarks: Demeritt (1973, 163).

Cavanugh & Irvin**   Robbinston   -1855-
Tools Made: Pump and Block Makers
Remarks: They are listed in the 1855 *Maine Registry and Business Directory*.

Cellis, Ira   Berwick   -1879-1890
Tools Made: Axes, Edge Tools, and Knives

Center Draft Mowing Machine Co.   Portland   -1879-1874
Tools Made: Farm Tools
Remarks: It is unclear if company made anything other than mowing machines.

Chaffin, R.**   Acton
Tools Made: Coopers' Tools
Remarks: Bob Jones has in his collection a coopers' howell marked **R.CHAFFIN. | ACTON.** with the "N" reversed. It is believed it was possibly made in Maine. The ink stamp and sketch is courtesy of Bob Jones.

Champion Axe Mfg. Co.**   Evart
Tools Made: Axes

Remarks: Champion is listed in Appendix 2 of Klenman's (1990, 99) *Axe Makers of North America* and is noted as an manufacturer of axes. Klenman does not give any further description of the Champion Axe Mfg. Co. in his text. Tom Lamond (personal communications) provides the following notes: The Champion name was used in the late 19th and early 20th centuries, in one context as a brand used by the Kelly Axe & Tool Co. of Charleston, W. VA. The Champion name was also used by various hardware and tool distributors. There was a Champion Tool Co. in Toledo, Ohio from 1922-1937 but I have never come across any specific reference to a Champion Axe Company other than when used as a non de plume by the Mann Edge Tool Co. of Lewistown, PA. It appears that the name "Champion" was used in a number of situations where it was actually as a brand, possibly a "house" brand. This made it easier for companies to sell to merchants that did not choose to sell the same brands as their local competitors, i.e. same product under a different name. The reference to Evart probably applies to The Evart Logging Tool Company which was located in Evart, Michigan.

Chandler, Charles H.**   East Machias   -1856-
Tools Made: Shipsmith
Remarks: He is listed in the 1856 *Maine Business Directory*.

Chandler, John   Stetson   -1849-
Tools Made: Edge Tools

Chandler, Moses   East Corinth   1869-1885-
Tools Made: Farm Tools, including Harrows, Cultivators, and Horseshoes
Remarks: His location is sometimes noted as Corinth in directories. He is credited for inventing the horse-hoe, according to the Corinth Historical Society.

Chaplin, Charles C.**   Liberty   -1870-
Tools Made: Machinist
Remarks: He is listed in the 1870 census.

Chapman, T. M., & Bros.   Old Town   -1869-
Tools Made: Saw Tools, including Saw Filing Machines

Chapman, Job**   Bath   -1855-
Tools Made: Shipsmith
Remarks: He is noted in Baker's (1973) *Maritime History of Bath*.

Chase, G., & Co.   Portland   -1841-1846-
Tools Made: Wooden Planes
Remarks: George Chase was a joiner and ships' carpenter. The imprint **G CHASE** has

been found on a 13 3/8" sash plane with two irons, an ovolo, and a rabbet, all made of beech (Pollak 2001, 85). The **G. CHASE & CO.** mark "has been reported on a 9 13/16" long beech rabbet and a 9 7/16" round" (Pollak 2001, 86). The Bob Jones collection has a 9 3/8" moving filetster marked **G. CHASE** and a 9 1/8" **G. CHASE & Co.** skewed rabbet. DATM (Nelson 1999) also lists a George Chase as a maker of whetstones, 1888-1898, location unknown.

Chase, C.
Tools Made: Wood Planes
Remarks: "All known planes with this mark were found in the vicinity of Portland, Maine" (Nelson, 1999). "Several planes have been found in the Portland, ME, area, including a 9 13/16" molder and a 9 ¾" quarter round, both with wide rounded chamfers, and a double handled 14 ½" match plane with metal skate and runner" (Pollak 2001, 85).

Chase, Charles F.   Dixfield   -1849-
Tools Made: Farm Tools

Chase, George**   -1841-1846-
Tools Made: Planes
Remarks: See G. Chase & Co.

Chase, Timothy**   Belfast   -1840-1875   (d. 1875)
Tools Made: Gunsmith, Jeweler, Machinist, and Clockmaker
Remarks: Demeritt (1973, 163).

Chase, William**   Frankfort   -1820-
Tools Made: Blacksmith
Remarks: When he was four years old, his parents moved to Frankfort to build a log house. William learned the trade of blacksmith, which he followed in connection with farming, and, in his day he was considered one of the best ox-shoers in the state. He served in the war of 1812 (*Biographical Review* 1897, 336).

Chase Turbine Mfg. Co.   Portland   -1886-
Tools Made: Circular Saws
Remarks: This company is probably related to the Chase Manufacturing Company of Orange, MA.

Chick, George   Portland, Bath   -1850-
Tools Made: Wood Planes
Remarks: Pollak (2001, 87) notes that there is one example marked **BATH. MAINE**, a 22" fore razee made of lignum vitae with a closed tote and round chamfers located in the

Bob Jones collection. The Jones collection also has an 8 ½" smooth plane made in Portland. DATM (Nelson 1999) notes that Chick's planes are "shipsmith types." A Chick ebony miter plane was sold at the Brown Auction, October 29, 2005.

Chick, M. L.**   ?
Tools Made: Planes
Remarks: The Bob Jones collection has a 12 1/4" shipbuilders' rabbet squared plane marked **M.L.CHICK**, found and probably made in Maine. This maker is not listed in Pollak (2001) or DATM (Nelson 1999).

Chick, William E.   Bangor   -1899-1900-
Tools Made: Farm Tools

Church, Benjamin**   Somerset County   -1814-
Tools Made: Blacksmith
Remarks: He is listed as owning 2 guns valued at $4.50. Source: Somerset County Probate Inventories

Church, Randell**   Bath   1867-1872
Tools Made: Gunsmith
Remarks: Demeritt (1973, 163).

Churchill, Bela**   Buckfield   -1820-1860-
Tools Made: Gunsmith
Remarks: Demeritt (1973, 163).

Clapp, Galen**   Bath   -1855-
Tools Made: Brass Founders and Finishers
Remarks: He is listed in the 1855 *Maine Registry and Business Directory*.

Clark & Edgerly**   Biddeford   -1870-1880-
Tools Made: Gunsmith
Remarks: Demeritt (1973, 163) notes this is Charles B. Clark and Samuel H. Edgerly.

Clark & Parsons Co.   East Wilton   1894-1913-
Tools Made: Farm Tools, including Hay Knives, Scythes, Machetes, Clippers, Corn Knives, Bread Knives, and Sickles
Remarks: Franklin J. Clark and Arthur D. Parsons used the brand name **Blue Ribbon**. According to W. A. "Chet" Sweatt's research, Hiram Holt & Co. was reorganized to form this company in 1894. E. J. Clark of Farmington was the president and A. Parsons of E. Wilton was the treasurer. They sold out to the Dunn Edge Tool Co. in 1904. A 25" cane

knife with a horn handle in a private collection in E. Wilton is marked **CLARK & PARSONS CO. | EAST WILTON, ME. U.S.A. | ACERO FINO | CALIDAD | GARANTIZADA | RELAMPAGO** (sideways) | **NO. 32** (sideways) with a touchmark of an arm holding a cane knife.

Clark, George H.**   Skowhegan   1887
Tools Made: Tool for Clinching Horse Nails
Remarks: He has Pat. No. 365,974, July 5, 1887, a tool for clinching horse-nails. One half of the patent is assigned to John W. Delano of the same place (Tom Lamond personal communication).

Clark, James M.**   Baldwin   -1855-
Tools Made: Pump and Block Makers
Remarks: He is listed in the 1855 *Maine Registry and Business Directory*.

Cleaves, Charles J.**   Biddeford   1880-1889
Tools Made: Gunsmith and Jeweler
Remarks: Demeritt (1973, 163).

Cleaves, B. F.**   Machias   -1855-
Tools Made: Wheelwright
Remarks: He is listed in the 1855 *Maine Registry and Business Directory*.

Clement, A.**   ?   -1850-
Tools Made: Planes
Remarks: Pollak (2001, 91) mentions a 22" beech razee fore plane that was found in Maine and is now in the Bob Jones Collection.

Clement, Albion**   Montville   -1870-1880-
Tools Made: Blacksmith
Remarks: He was the son of William Clement and is listed in the 1870 and 1880 census.

Clement, William**   Montville   -1850-1860-
Tools Made: Blacksmith
Remarks: He is listed in the 1850 and 1860 census and was born in Montville.

Clemons, William H.**   Hiram   -1860-1870-
Tools Made: Gunsmith
Remarks: Demeritt (1973, 163).

Clifford, J. K.   Westbrook   -1856-
Tools Made: Axes

Clifford, Timothy K.**   Denmark   -1855-
Tools Made: Rakes
Remarks: He is listed in the 1855 *Maine Registry and Business Directory.*

Close, A. M.**   Calais   -1855-
Tools Made: Wheelwright
Remarks: He is listed in the 1855 *Maine Registry and Business Directory.*

Cluff, J. L.**   Skowhegan   ca. 1850
Tools Made: Planes
Remarks: A fruitwood block plane was found in the Skowhegan area and a 22" razee ship fore plane has also been reported (Pollak 2001, 92). An 11" birch tongue and groove plane is in the Bob Jones collection.

Cobb, Lyman H.**   South Portland   1899
Tools Made: Gunsmith
Remarks: Demeritt (1973, 159-60, 163) notes his 1899 patented shotgun was sold by the John P. Lovell Arms Company, also of S. Portland.

Coffin, G. W.   Freeport
Tools Made: Plane Irons and Wood Planes
Remarks: DATM (Nelson 1999) indicates that Coffin was a ships' carpenter in Freeport, date unknown. Pollak (2001, 93) lists a 16 ½" razee beech crown molder, marked **G.W.COFFIN**, with shallow round chamfers and Moulson Brothers iron, ca.1840. There is a possible alternate spelling of Coffen.

Cofran, William   Readfield   -1871-
Tools Made: Farm Tools, including Cultivators

Cogan, James   Houlton   -1899-1900-
Tools Made: Farm Tools

Cole, H. M.   Hope   -1879-1885-
Tools Made: Farm Tools, including Mowers.

Collett, F. & J.   Bangor   -1847-1849-
Tools Made: Files
Remarks: Also listed as T. & J. Collett.

Collett, Job   Bangor   -1855-1894-
  Tools Made: Files
  Remarks: "Probably a part of F. & J. Collet ... in 1855 his name was linked to Michael Schwartz, but their working name is not known" (Nelson, 1999).

Collins, J. W., & Co.**   Montville   -1881-1882-
  Tools Made: Axes
  Remarks: There is a J. W. Collins & Co. located in Montville that made axes listed in the 1881 and 1882 *Maine Business Directories*. It is unknown if this is W. Collins of Liberty or someone else.

Collins, W.*   Liberty, Montville   1881-1884
  Tools Made: Axes
  Remarks: Collins worked in Liberty until 1882, then moved to Montville where he continued until 1884 (Yeaton 2000).

Colomy, Isaac, Jr.**   Somerville   -1867-
  Tools Made: Axes, Shovels, Chisels, etc.
  Remarks: He is listed in the 1867 *Maine Business Directory*.

Colville, Alexander A.**   Little River (Berwick)   -1820-
  Tools Made: Gunsmith
  Remarks: Demeritt (1973, 29, 163) notes he had a patent for a gun lock.

Cook, Joseph**   Porter   -1860-1870-
  Tools Made: Gunsmith
  Remarks: Demeritt (1973, 163).

Conant, C. M., Co.   Bangor   -1899-1900-
  Tools Made: Farm Tools

Conelon, Charles R.   East Machias   -1856-
  Tools Made: Edge Tools
  Remarks: He is listed in the 1856 *Maine Business Directory* as an edge toolmaker.

Connor, John, & Son **   Portland   -1855-
  Tools Made: Pump and Block Makers
  Remarks: They are listed in the 1855 *Maine Registry and Business Directory*.

Coombs, B. D.**
  Tools Made: Planes

Remarks: A rounding plane marked **B.D.COOMBS.** in the Bob Jones collection was located in Maine. It is not know if there was any connection with J. H. Coombs of Lisbon Falls.

Coombs, I.** Bangor
Tools Made: Planes
Remarks: The Bob Jones collection contains a 9 ½" skewed rabbet plane marked **I.COOMBS | BANGOR | W.SEWARD.**

Coombs, John H. Lisbon Falls -1885-
Tools Made: Farm Tools

Coombs, L. A** Vinalhaven
Remarks: Pollak (2001, 102) states that L. A. Coombs was possibly a hardware dealer in Vinalhaven, ME. The example listed is a 1 ¼" nosing plane marked **L.A.COOMBS | VINAL HAVEN | ME** made by the Ohio Tool Co.

Cooper, Andrew** Sedgewick -1856-
Tools Made: Shipsmith
Remarks: He is listed in the 1856 *Maine Business Directory*.

Copp, Dunton* Liberty 1856-
Tools Made: Axes

Cord, Francis** Robbinston -1856-
Tools Made: Shipsmith
Remarks: He is listed in the 1856 *Maine Business Directory*.

Corliss, William E.** Bath -1869-1882-
Tools Made: Shipsmith
Remarks: He is listed in the 1869, 1881, and 1882 *Maine Business Directories* at Broad St.

Cotton, H. P. Damariscotta Mills -1879-
Tools Made: Farm Tools and Rakes, including Mowing Machines and Horse Rakes

Cram, Danvers** Liberty 1891
Tools Made: Wrenches
Remarks: "Inventor and maker of an adjustable axle nut wrench, patented [445,258] January 27, 1891" (Cope, 1999). It is illustrated in an

advertisement and the only specimen known was recently purchased in an auction (photo).

Cram, Elija**   Liberty   -1850-
Tools Made: Cram Water Wheel
Remarks: "The Cram water wheel was developed around 1850 by Elijah, son of Jesse Cram. It sat horizontally, turning a vertical shaft. ...The wood bearing was said to withstand several years of continuous use. ...The wheels were generally housed in a wooden enclosure called a tub or wheel case that improved efficiency by preventing the water from escaping sideways. Cram wheels were common in the mills of the area for many years" (Donahue 1996).

Cram, Ira**   Liberty   -1856-1870-   (b. August 26, 1837)
Tools Made: Blacksmith and Machinist
Remarks: He was the son of Jesse and Martha (Dutton) Cram and the grandson of Smith Cram. He worked as a farmer, nurseryman, and merchant; owned a machine and blacksmith shop, lumber mill, cider mill, and tree nursery; and sold threshing machinery and building materials (Donahue 1996).

Cram, L., & Co. **   Bangor   -1840-
Tools Made: Planes
Remarks: Pollak (2001, 106) lists two planes with the **L. CRAM & Co** mark. The second mark also states **BANGOR**. The Bob Jones collection has a 9 ½" Grecian ovolo plane with the first mark that is also stamped **NSW**.

Creamer, Cyrus I.   Nobleboro   -1879-
Tools Made: Farm Tools, including Mowing Machines, Horse Rakes, Plows, and Cultivators

Creasey, Isaiah**   Montville   -1880-
Tools Made: Machinist
Remarks: He is listed in the 1880 census.

Creasey, Isaiah E.**   Montville   -1880-
Tools Made: Machinist
Remarks: He is listed in the 1880 census. Son of Isaiah Creasey.

Creighton & Co.   Union   -1867-1871-
Tools Made: Axes
Remarks: This company is listed in the 1867 and 1869 *Maine Business Directories.*

Crie, H. H., & Co.   Rockland   1860-1914

Tools Made: Planes

Remarks: H. H. Crie and R. Anson Crie established themselves in 1860 as hardware dealers in Rockland, ME, and were listed in business directories until 1914. Their imprint **H.H. CRIE & CO.** has been seen stamped on the edge and side of a croze made by D.B. Titus (w.s.) (Pollak 2001, 109). The Bob Jones collection has a 17 ½" drawshave with this imprint and **L.&I.J.WHITE | BUFFALO,NY**. It's probable that the Crie company did not make these tools, but rather were their purveyors to local woodworkers and fishermen. The Davistown Museum has a Dickinson swivel (ID# 112004T1), which is a hand line weight for cod fishing, also known as a George's Bank sinker, stamped **H.H. CRIE & CO. ROCKLAND ME** in the IR collection. Numerous cod fishing lead sinkers with the Crie stamp have been observed at the Liberty Tool Co.

Crockett, George B.   West Sumner   -1871-

Tools Made: Farm Tools, including Drag Rakes

Crockett, George W.   Gorham   -1869-

Tools Made: Farm Tools, including Mowing Machines

Crockett, Leonard**   Portland   -1855-

Tools Made: Brass Founders and Finishers

Remarks: He is listed in the 1855 *Maine Registry and Business Directory*.

Crogan Mfg. Co.   Bangor   -1915-1917-

Tools Made: Rules

Remarks: The Davistown Museum has a 100 foot and 25 foot steel tape measure (ID# 5100T9) in the IR collection. Numerous

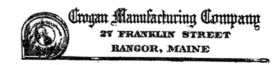

examples of Crogan tape measures have been noted. The figure of the return address from an envelope with a 2 cent stamp is courtesy of Raymond Strout.

Crompton, Isaiah**   Calais   -1856-

Tools Made: Shipsmith

Remarks: He is listed in the 1856 *Maine Business Directory*.

Cromwell, Simon**   Edgecomb   1827

Tools Made: Gunsmith

Remarks: Demeritt (1973, 31, 163) notes he had a patent for a gun lock.

Crooker & Bartlett   Foxcroft   -1849-
Tools Made: Hoes and Forks
Remarks: They are listed in the 1855 *Maine Registry and Business Directory*.

Crooker & Lilly**   Bath   1855-1882-
Tools Made: Shipsmith
Remarks: They are listed in the 1881 and 1882 *Maine Business Directories* on Commercial St. See David Crooker and Robert Lilly.

Crooker, David**   Bath   -1840-1882-
Tools Made: Shipsmith
Remarks: In 1840, David Crooker began "forging of anchors, capstans, windlass necks, and other heavy iron work for ships... Crooker and [Robert] Lilly became partners in 1855; all their forging was by hand until 1865 when they installed a small steam hammer" (Baker 1973, 434).

Crooker, Isaiah**   Bath   1750-1759
Tools Made: Blacksmith
Remarks: He forged many of the handmade nails that would build the early homes along the Kennebec. He served as a Private in the Paris Militia Company during the American Revolution. This information provided by Barbara Ann Crooker, Box 631, Bath, ME 04530.

Crooker, Jonathan Harding**   Bath   1785-1805
Tools Made: Blacksmith and Shipbuilder
Remarks: Jonathan was the second son of Isaiah Crooker and he learned the trade from Isaiah. This information provided by Barbara Ann Crooker, Box 631, Bath, ME 04530.

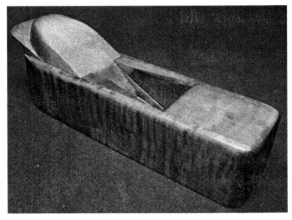

Crown Plane Co.**   Bath, S. Portland   -1985-2001

Tools Made: Planes

Remarks: The Crown Plane Co. was originally in Bath and owned by Leon Robbins. In 2001, he sold it to James White of South Portland, who uses a **J W** mark with a crown (photo). Numerous planes made by Robbins are now in the Davistown Museum collection.

Cummings, Leonard F.   Porter, Gray   -1849-1871-

Tools Made: Axes and Edge Tools

Remarks: Cummings moved from Porter to Gray sometime between 1857 and 1858 (DATM, 1999). He is listed as an edge toolmaker in Porter in the 1855 and 1856 *Maine Business Directories* and as an ax-maker in Gray in the 1867 and 1869 *Maine Business Directories.*

Currier, Jonathan**   Waldoboro   -1855-

Tools Made: Wheelwright

Remarks: He is listed in the 1855 *Maine Registry and Business Directory.*

Currier, Micajah   East Orrington   -1869-1871

Tools Made: Farm Tools, including Hay Rakes

Currier, W.**   -1800-

Tools Made: Planes

Remarks: Pollak (2001, 111) lists a 10 ½" Yankee plow marked **W. CURRIER** with square slide-arms stopped with both wedges and wood thumb screws, no depth stop, and flat chamfers, that was found in Maine.

Curtis, A. J.**   Monroe

Tools Made: Screwdrivers

Remarks: Curtis obtained patent 192901, 7/10/1877, information courtesy of C.D. Fales.

Curtis, Ezra**   Blue Hill   -1855-

Tools Made: Wheelwright

Remarks: He is listed in the 1855 *Maine Registry and Business Directory.*

Curtis, Nehemiah**   Harpswell   1700s

Tools Made: Blacksmith

Remarks: He is the first blacksmith in Harpswell for which there is any account (Wheeler 1878).

Curtis, Nehemiah**   Harpswell   late 1700s
Tools Made: Blacksmith
Remarks: The son of Nehemiah Curtis, he assumed his father's business (Wheeler 1878).

Curtis, Nehemiah**   Harpswell   -1820-
Tools Made: Blacksmith
Remarks: The grandson of Nehemiah Curtis, he assumed his father's business (Wheeler 1878).

Cushing, A. P.**   East Machias   -1856-
Tools Made: Shipsmith
Remarks: He is listed in the 1856 *Maine Business Directory*.

Cushing, Royal J.**   Boothbay   1884
Tools Made: Gunsmith
Remarks: Demeritt (1973, 163).

Cushing, Ruel J.**   Bangor   1881-1884
Tools Made: Gunsmith
Remarks: Demeritt (1973, 163).

Cushman, Henry R.   Andover, South Andover   -1855-1879-
Tools Made: Edge Tools
Remarks: He is listed in the 1855 and 1856 *Maine Business Directories* in Andover, ME, and is listed in the 1879 *Maine Business Directory* in both Andover and South Andover.

Cushman, William Gregg**   Andover   -1885-   (b. Oct. 26, 1852)
Tools Made: Log Calipers
Remarks: A log caliper (photo) was found in Nova Scotia signed **Manufactured | By | W. G. CUSHMAN, | Andover, Me.** Research by the discoverer of the tool determined that W. G. Cushman was the son of Henry R. Cushman and Barbara Gregg Cushman of Andover (*Maine Births and Deaths* Book 1795 - 1870, 40).

Cutler, Levi   Kingsbury   -1856-
Tools Made: Edge Tools
Remarks: He is listed in the 1856 *Maine Business Directory* as an edge toolmaker.

Cutts, Samuel   North New Portland, Pittston   -1849-1871-

Tools Made: Edge Tools and Plows

Remarks: There may have been two different men with the same name, one operating in North New Portland making edge tools in 1849 and another Samuel Cutts making plows in Pittston in 1871.

Cyphers, M. B.**   Skowhegan   -1860-

Tools Made: Gunsmith

Remarks: Demeritt (1973, 163) notes he worked in Greenville, Michigan in the 1870s to 1900.

D. H.**

Tools Made: Planes

Remarks: "More than 30 examples have been reported bearing this imprint, most from the Auburn to Skowhegan section of Maine. The initialed imprint is similar to that used by L. Sampson (w.s.). Appearance is ca. 1800" (Pollak 2001, 180). The Bob Jones collection has two examples of planes with the mark **D H**; between the two letters is an abstract evergreen tree.

Daggett, Cyrus**   Sherman   -1875-1879-

Tools Made: Axes and Gunsmith

Remarks: He is listed in the 1879 *Maine Business Directory*. The drawshave in the photograph is privately owned and marked **C. DAGGETT**. The Davistown Museum has recently acquired a second C. Daggett drawshave slightly larger than the one shown (15 ¾" long with 9 7/8" long cutting blade). Demeritt (1973, 163) lists a C. Daggett as a gunsmith in Sherman in 1875.

Damon Brothers*   Carmel, Oakland   1904-1928-

Tools Made: Axes

Remarks: Klenman (1990) lists Damon Brothers as an Oakland, Maine ax-maker, one of the many ax-making companies located along the Emerson Stream going into Waterville. Yeaton (2000) notes that Emerson & Stevens Mfg. Co. of Oakland bought them out in 1928.

Daniels, Zachariah**   Newfield, Pittston   -1867-1871-

Tools Made: Axes

Remarks: He is listed in the 1867 and 1869 *Maine Business Directory* as making axes in Newfield. Newfield is near Wolfeboro (greater Portland area). Zachariah Daniels is also listed in DATM (Nelson 1999) as working in Pittston (near Gardiner) from 1869 -

1871 making axes and edge tools. It is unknown if he moved or if this is two different people.

Darling & Bailey   Bangor   1852-1853
Tools Made: Rules
Remarks: The Davistown Museum has a steel rule (ID# 71903T6) marked **D. & B. | Bangor Me**, in the MIV collection. In 1853, Samuel Darling bought out Edward H. Bailey and with a new partner became Darling & Schwartz. Darling & Bailey is a very rare maker's signature from the very beginning of the classic period of American machinist tools.

Darling & Schwartz   Bangor   1854-1866
Tools Made: Machinist Tools, Metal Planes, Rules, and Squares
Remarks: Most of their rules are marked **D. & S. BANGOR**. DATM (Nelson 1999) notes following: "Samuel Darling and Michael Schwartz (who succeeded Darling & Bailey) made try squares with 1852 (possibly Nathan Ames' 6 July 1852) and 6 Oct. 1857 (Darling) patents and circular iron planes patented by George F. Evans in 1862 and 1864. The 1857 patent square was later made by Darling, Brown & Sharpe after Darling joined J.R. Brown & Sharpe in 1866.

Schwartz worked otherwise as a Bangor saw-maker and hardware dealer and did not join the new company." This is another important company from the early days of the classic period of American machinist tools. The Davistown Museum has a D & S rule (ID# 42801T5) and a depth gauge (ID# 41203T5) in the MIV collection (photo).

Darling, Samuel   Bangor
Tools Made: Machinist Tools
Remarks: Although Darling was part of several companies at different times (Darling & Bailey and Darling & Schwartz in Maine, then in 1866, Darling Brown & Sharpe in Rhode Island), it is believed that at one time, either before joining with Bailey or between companies, he worked independently. He had three patents, one for a straight edge (7 Jan. 1868), a vise (7 April 1868), and a file hardening process (12 July 1870).

Davenport, Anthony   Portland
Tools Made: Scientific Instruments, including survey compasses

Davies, Frank   Sidney   -1879-
Tools Made: Farm Tools, including pumps, harrows, and hand sleds
Remarks: Also cited as working in North Sidney.

Davis & Blake   Portland   c.1851-
Tools Made: Box Openers
Remarks: The photograph is of a box opener marked **DAVIS & BLAKE | MANDFRS. PORTLAND. ME. | PATENTED | OCT. 21, 1851.** The patent number is 8457 and is held by Geo. C. Taft of Worcester, Massachusetts. Thank you to Rick Floyd for the box opener information and photograph. The DATM (Nelson 1999, 213) listing for this company includes a reference to a floor laying clamp with a 21 Oct. 1851 patent date. We think that DATM has mis-identified the name/use of this tool, as this photo matches the patent office drawing.

Davis & Clark   Portland   c.1851
Tools Made: Clamps
Remarks: Also made a floor laying clamp with a 21 Oct. 1851 patent date. See Davis & Blake.

Davis, Daniel   Freeman   -1856-
Tools Made: Axes and Edge Tools
Remarks: He is listed in the 1856 *Maine Business Directory* as an edge toolmaker.

Davis, Dexter**   Montville   -1880-
Tools Made: Blacksmith
Remarks: He is listed in the 1880 census.

Davis, Ezra   Long Island   -1856-
Tools Made: Edge Tools
Remarks: He is listed in the 1856 *Maine Business Directory*.

Davis, H. R.   West Paris
Tools Made: Chisels, including Barking Spuds
Remarks: The initial H. is stated as an M. in one source (Nelson 1999).

Davis, Oliver**   Kennebunkport   -1855-
Tools Made: Pump and Block Makers
Remarks: He is listed in the 1855 *Maine Registry and Business Directory*.

Davis, Samuel   Worcester   -1840-
Tools Made: Wool Spinning Machines

Davis, Sylvanus**   Lincoln   -1870-

Tools Made: Gunsmith
Remarks: Demeritt (1973, 163).

Davis, T. B. Arms Co.**   Portland   -1870-1900-
Tools Made: Gunsmith
Remarks: Demeritt (1973, 91-2, 163) indicates this is Theodore Davis.

Davis, William H.**   Linneus   -1875-
Tools Made: Gunsmith
Remarks: Demeritt (1973, 163).

Day, Benjamin**   Cherryfield, Bangor   -1856-1859-
Tools Made: Gunsmith
Remarks: Demeritt (1973, 163) notes him in Cherryfield in 1856 and Bangor in 1859.

Day, G.**   1810-
Tools Made: Planes
Remarks: Pollak (2001, 116) lists a skew rabbet and a complex molder by Collins of Hartford, marked **G.DAY**, both 9 1/4" with flat chamfers. Also reported is a 16 1/2" apple wood razee shipwright's jack with flat chamfers, which was found in Maine.

Dearborn, J.**   -1790-1830-
Tools Made: Planes
Remarks: Pollak (2001, 118) states that the **A** imprint and **J D** initial group is on a 9 3/4" beech complex molder with heavy flat chamfers, ca. 1790. The **A** imprint alone is possibly from a second generation and is on a 9 1/4" single-boxed beech molder, a double-wedge/iron fixed sash, found in Maine, and a 9 1/8" fruitwood slide-arm plow with thumbscrews, brass arm tips, brass depth stop, and brass plate on the skate, and heavy round chamfers. These are marked **J D** (early) or **J.DEARBORN** (later).

Dearborn, Otis R.**   Limerick   1871
Tools Made: Gunsmith
Remarks: Demeritt (1973, 163).

Deering, J. R.   Saco   -1849-1881-
Tools Made: Edge Tools
Remarks: He is listed in the 1881 *Maine Business Directory* as located opposite the Eastern Railroad Depot. Will Hight of Freeport has a 2" J. R. Deering framing chisel.

Delano, John W.**   Skowhegan   1887
Tools Made: Tool for Clinching Horse Nails

Remarks: He has Pat. No. 365,974, a tool for clinching horse-nails, July 5, 1887. One half of the patent is assigned to George H. Clark of the same place (Tom Lamond personal communication).

Delanti, C. L.   Ellsworth   -1855-1856-
Tools Made: Edge Tools
Remarks: He is listed in the 1855 and 1856 *Maine Business Directories* as an edge toolmaker.

Deming**   Calais   1885
Tools Made: Line Rule

Deming, Frank M.**   Flagstaff   1906
Tools Made: Gunsmith
Remarks: Demeritt (1973, 163).

Dennett, Samuel   Kittery   -1744-
Tools Made: Blacksmith, Chisels, and Caulking Irons

Dennis, Hazen**   Liberty   -1870-1880-
Tools Made: Blacksmith
Remarks: He is listed in the 1870 and 1880 census.

Dennis, J. M.   East New Portland   -1855-1856-
Tools Made: Axes and Edge Tools
Remarks: The 1855 *Maine Business Directory* lists I. M. Dennis of New Portland, probably a mistype of his name. He is listed in the 1856 *Maine Business*

Directory as J. M. Dennis of New Portland. The Davistown Museum has a drawknife donated by Roger Smith (ID# 30801T2) marked **J M DENNIS EAST NEW-PORTLAND** in the MIV collection.

Dennison, Charles   Freeport, Farmington   -1867-1871
Tools Made: Axes and Wood Planes
Remarks: An ax-maker by the name of Charles Dennison working around the dates 1869-1871 was listed in the 1867 and 1869 *Maine Business Directories*. A plane marked **C.H. DENNISON | FREEPORT ME.** has been recovered and is said to have been made by a Charles Dennison (born 1835), who was a joiner in 1860 and a carpenter in 1895 and 1901 (Nelson 1999). Yeaton (2000) lists two Charles Dennisons, both ax-makers, one in

Farmington, the other in Freeport. The Bob Jones collection has a 7" slightly rounded lignum vitae smooth plane with this mark. Pollak (2001, 121) also lists Dennison.

Dennison, J. M.**   East Machias   -1856-
Tools Made: Shipsmith
Remarks: He is listed in the 1856 *Maine Business Directory*.

Dennison, Joseph W.**   Freeport   1871
Tools Made: Gunsmith
Remarks: Demeritt (1973, 163).

Deshon, S. M.**   Kennebunkport   -1881-
Tools Made: Edge Tools
Remarks: He is listed in the 1881 *Maine Business Directory*.

Devereaux, S. K.**   Castine   -1856-
Tools Made: Shipsmith
Remarks: He is listed in the 1856 *Maine Business Directory*.

Dewolf, Isaac & Co.   Portland   -1849-
Tools Made: Farm Tools and Edge Tools
Remarks: An Issaac Dewolf cast steel slick (3 1/3" wide) was located in a Camden, Maine, tool chest and sold by the Hulls Cove Tool Barn to Tom Lunford of Lake Junction, Texas, on Sept. 20, 2003.

Diamond Wrench Mfg. Co.   Portland   -1880-1900-
Tools Made: Wrenches
Remarks: Their signature was **DIAMOND WRENCH | STEEL FORGED | PORTLAND, ME.** with a diamond shape figure with the patent date in it. The two patents used were a 16 Oct. 1883 patent and a 2 Nov. 1880 patent belonging to Henry A. Thompson of Farmington, ME, but it is unknown if he was involved with the company. The two wrenches in the photo are owned by Herb Page; 18 ½" and

5 ½" combination nut and buggy wrenches. The larger is stamped **DIAMOND WRENCH CO. PORTLAND, ME** and the smaller **DIAMOND WRENCH MFG. CO. PORTLAND, ME.** Both have the two patent dates stamped on them. The second photo is a close up of yet another maker's mark on a wrench on loan to the Davistown Museum.

Dillingham, J.** Turner -1862-
Tools Made: Handsaws
Remarks: He had patent 34946 on 4/15/1862 for a handsaw. Source: Graham Stubbs.

Dillingham, T. H. Old Town -1855-
Tools Made: Farm Tools

Dingee & Mosher Presque Isle -1856-
Tools Made: Axes and Edge Tools
Remarks: This company is listed in the 1856 *Maine Business Directory*.

Dingley Brothers** Gardiner -1867-
Tools Made: Edge Tools
Remarks: This company is listed in the 1867 *Maine Business Directory*.

Dirigo Saw Works Bangor -1869-1877-
Tools Made: Saws
Remarks: From 1869-1870 this company is listed at the same address as Gibson, Kimball & Sanford and Kimball & Sanford; it is thought that this may have been the name of their factory. A jigsaw bearing a J. W. Penney 3 July 1877 patent also has the name "Dirigo" (without Saw Works), "Mechanic Falls, ME" on it. This could be the mark of another maker (Nelson 1999). See Penney & Thurston, also of Mechanics Falls.

Doble, Benjamin W. Lagrange -1856-1881-
Tools Made: Axes
Remarks: He is listed in the 1856, 1867, 1869, 1879, and 1881 *Maine Business Directories*. The 1856 Directory lists him as B. W. Doble under edge toolmakers.

Doble, B. W., Jr. Milo -1881-1907- (1855-1912)
Tools Made: Axes
Remarks: He is listed in the 1881 and 1882 *Maine Business Directory*. The 1907 Milo Business Directory lists him as a blacksmith. His gravestone is in the Evergreen Cemetery in Milo.

Doble, W.* Milo 1882-1925
Tools Made: Axes

Doble, William Jr.** Milo -1879-1907-
Tools Made: Edge Tools
Remarks: He is listed in the 1879 *Maine Business Directory*. The 1907 *Milo Business Directory* lists W. B. Doble Jr. as a hardware dealer. It is unknown if this is the ax-maker

that Donald Yeaton (2000) listed as W. Doble or if the Dobles are all relatives. The towns of Milo and LaGrange are very close to each other.

Doe, Hiram　Vassalboro　-1855-1885-
Tools Made: Plows
Remarks: Also listed in North Vassalboro.

Dolbier, D. C.*　Kingfield　1875-1885
Tools Made: Axes

Dolbier, W. S. & W.　Kingfield　-1867-1871-
Tools Made: Axes
Remarks: This company is listed in the 1867 and 1869 *Maine Business Directory*. Wm. Dolbier of Kingfield is listed in the 1882 *Maine Business Directory* as an ax-maker.

Dolbier, William　Kingfield　-1862-1879-
Tools Made: Axes and Edge Tools
Remarks: He also worked on carriages and may have made them. He is listed in the 1879 *Maine Business Directory*. A coopers' adz (photo) has been found that is marked **W. DOLBIER**.

Donnell, C. A. & Co. **　Portland　-1855-
Tools Made: Brass Founders and Finishers
Remarks: They are listed in the 1855 *Maine Registry and Business Directory*.

Doten, T.**　-1800-
Tools Made: Planes
Remarks: Pollak (2001, 127) lists a double-boxed, complex molder and a crown molder with a cherry wedge and offset tote, both beech with flat chamfers, found in Maine. Both are marked **T.DOTEN**.

Douglass, Luther**　South Bridgton, Bridgton　-1881-
Tools Made: Axes
Remarks: He is listed in the 1881 *Maine Business Directory*.

Dow, Oliver　Dayton　-1856-1871-
Tools Made: Axes
Remarks: He is listed in the 1856 and 1869 *Maine Business Directories* as an edge toolmaker.

84

Dow, W. N.**   Head Tide   ca. 1850
Tools Made: Planes
Remarks: Pollak (2001, 128) lists the 8" coffin smooth plane in the Bob Jones collection that has the mark **W. N. DOW | HEAD TIDE ME.**

Doyen, Samuel   Bangor   -1867-1875-
Tools Made: Cooper Tools
Remarks: Doyen also worked as a cooper. Pollak (2001, 128) states that Samuel Doyen advertised in 1867 in the *Maine Business Directory* as a "Manufacturer of coopers tools" in Bangor, ME from 1867-1875. Matthew Moriarty may have apprenticed to him in 1867.

Drake, P.**   Appleton   -1879-
Tools Made: Axes
Remarks: He is listed in the 1879 *Maine Business Directory*.

Drake & Jones**   Union   -1855-
Tools Made: Shovels and Spades
Remarks: They are listed in the 1855 *Maine Registry and Business Directory*.

Drew, Israel**   Solon   1880-1881
Tools Made: Smith
Remarks: He is listed in Hoyt (1881).

Drew, Moses*   New Limerick   1878-1895
Tools Made: Axes, Chisels, etc.
Remarks: He is listed in the 1879 and 1882 *Maine Business Directories*.

Dumont, Worthington**   Biddeford   1856
Tools Made: Gunsmith
Remarks: Demeritt (1973, 163).

Duncan & Davenport**   Bath   -1855-
Tools Made: Pump and Block Makers
Remarks: They are listed in the 1855 *Maine Registry and Business Directory*. Baker (1973, 436) indicates they "...held the New England rights for the manufacture of Waterman & Russell's patent iron-strapped blocks."

Dunham, Daniel M. & Co.   Bangor   -1871-1872-
Tools Made: Hoes and Household Tools, including Looms and a Chandler Horse Hoe.

Dunn Edge Tool Co.   Oakland   -1858-1969-

Tools Made: Axes, Farm Tools, Scythes, and Sickles, including Hay Knives

Remarks: Using the two marks: **DUNN EDGE TOOL CO. | OAKLAND, ME.** and **D.E.T. CO. | OAKLAND ME.**, Reuben B. Dunn worked and lived in West Waterville, Oakland, and Fayette concurrently. Formerly of Dunn, Jordan & Co., he bought and ran other companies. The company is listed in the 1867, 1869, 1879, 1881, and 1882 *Maine Business Directories*. Dunn was one of Maine's largest manufacturers of axes and farm tools. The company stamp can either be Oakland or Waterville. This confusion can be explained by the fact that Oakland was originally part of Waterville before becoming a separate town; many of Oakland's ax-makers are also listed as Waterville ax-makers for the same reason. Most of Oakland's ax and tool factories are located on a one mile stretch of what was called the Emerson Stream; in today's *Maine Atlas* this stream is now labeled Messalonskee Stream, which drains into the Kennebec River after going through downtown Waterville. Dana Phillippi has two corn knives (photo of their mark) manufactured by this company. Figure of the bill courtesy of Raymond Strout.

Dunn, Elden & Co.   Waterville   -1855-

Tools Made: Farm Tools

Dunn, Jordan & Co.   Waterville   -1856-1857

Tools Made: Axes and Scythes

Remarks: Reuben B. Dunn and William Jordan

Dunn, Reuben, B.   Wayne   -1846-1849-   (b.1802, d.1889)

Tools Made: Scythes

Remarks: An 1846 North Wayne report lists Dunn as working under his own name. However, another source shows that he owned the Wayne Scythe Mfg. Co. in South Wayne from 1840-1845, but in 1849 lists him as just operating it. He later went on to buy and form other companies leading up to Dunn Edge Tool Co. *A History of the North Wayne Tool Co.* has recently been published by the Wayne Historical Society (Kallop 2003). It gives a chronology of the various names used by this company. Klenman (1990) lists Ruben as starting operations in Oakland, Maine, and becoming the largest ax manufacturer in the area.

Dunnels & Roberts*   Newfields   -1874-1876
Tools Made: Axes

Dunning, C. H. & Co.   Bangor   -1855-
Tools Made: Farm Tools

Dunning, R. B. & Co.
Bangor   1835-1897-
Tools Made: Oilers and
Agricultural Implements
Remarks: They also were
plumbing supply dealers, using
the brand name **DUNCO**. An
1897 source dates them back to
1835, but in the earlier years
they may have dealt or made

different products. The 1855 *Maine Registry and Business Directory* lists R. B. & A. Dunning as making agricultural implements. 1896 bill of sale courtesy of Raymond Strout.

Dunning, Andrew**   Brunswick   -1717-
Tools Made: Blacksmith
Remarks: Settled at Maquoit in 1717 (Wheeler 1878, 578).

Dunton, Copp & Co.   Liberty   -1856-
Tools Made: Axes
Remarks: See Dunton, Isaac L.

Dunton, Isaac L.   Liberty   -1862-
Tools Made: Axes and Edge Tools
Remarks: He might have been part of Dunton, Copp & Co., ca. 1856.

Dunton, J. L.**   Belfast   -1867-
Tools Made: Axes
Remarks: He is listed in the 1867 *Maine Business Directory*.

Durgin, C. S.   Forks   -1856-
Tools Made: Edge Tools
Remarks: He is listed in the 1856 *Maine Business Directory* as an edge toolmaker.

Dusten, Nathaniel & Co.   Dexter   -1869-1880-
Tools Made: Plows

Dustin, Joseph**   Brunswick   -1820-
Tools Made: Blacksmith

Dykeman, W. A.   Houlton   -1899-1900-
Tools Made: Farm Tools

Eames, James   Newry   ca. 1835-
Tools Made: Scientific Instruments
Remarks: Made and patented (11 Feb. 1835) a survey compass.

Eastman, Robert**   Brunswick   1829
Tools Made: Gunsmith and Clockmaker
Remarks: Demeritt (1973, 163) indicates he holds a patent.

Eastman, S. P.**   -1850-
Tools Made: Planes
Remarks: Pollak (2001, 133) lists a jack plane found in Gardiner, Maine, as having the imprint **S.P. EASTMAN** marked three times in a triangle.

Eastman, Stillman   Bradford   -1867-1887-
Tools Made: Cutlery
Remarks: He is listed in the 1867 and 1856 *Maine Business Directories* under edge tools: cutlery. The term "cutlery" could have included such items as spoke shaves and drawknives in addition to kitchen cutlery. The 1882 *Maine Business Directory* lists him making axes and knives in both North Bradford and Bradford.

Easton, John**   Lincolnville   -1869-
Tools Made: Shipsmith
Remarks: He is listed in the 1869 *Maine Business Directory*.

Edgecomb, Edward F.**   Mechanic Falls   1878
Tools Made: Gunsmith
Remarks: Demeritt (1973, 163) indicates he holds a patent.

Edgerly, Bingham**   Greenfield   -1870-

Tools Made: Gunsmith
Remarks: Demeritt (1973, 163).

Edmunds, J. P.* Dixfield -1882-
Tools Made: Edge Tools
Remarks: He is listed in the 1882 *Maine Business Directory*.

Edward, J.* -1800-
Tools Made: Planes
Remarks: Pollak (2001, 135) lists a 9 3/16" birch skew rabbet found in No. Windham, Maine, signed **J. EDWARD**.

Eldridge Bros. Dexter -1885-
Tools Made: Farm Tools and Household Tools
Remarks: One item advertised was a "World's Fair Churn."

Elliot, John* Thomaston -1855-
Tools Made: Pump and Block Makers
Remarks: He is listed in the 1855 *Maine Registry and Business Directory*.

Ellis & Norton Co. Kingfield -1896-1904-
Tools Made: Axes and Cant Dogs
Remarks: The company founders were Edwin Ellis, Willie F. Norton, and H. Abel Hunewell.

Ellis Saw Co. West Waterville -1867-1871
Tools Made: Saws
Remarks: The company also had a location in Boston. A bucksaw they made had a patented hold-down which may have been patented by Ellis. Possibly this is the patent issued to Erastus Bates of Waterville, which he assigned to John Ellis of North Bridgewater, MA.

Ellis, Benson* Corinna 1871
Tools Made: Gunsmith
Remarks: Demeritt (1973, 163).

Emerson & Stevens Co. Oakland 1870-1977
Tools Made: Axes, Hatchets, and Scythes
Remarks: Founded by Luther D. Emerson, Joseph E. Stevens, George W. Stevens, Charles E. Folsom, and William R. Pinkham, the company marked their tools **EMERSON, STEVENS&CO. | WEST WATERVILLE, MAINE**. (The Town of West Waterville became

Oakland in 1877.) This company is listed in the 1879, 1881, and 1882 *Maine Business Directories*. Shortly after 1902, they became the Emerson & Stevens Mfg. Co. An ax label for **THE WETMORE AXE** has been found (figure) and the owner is interested in information on this ax. An ax bearing the mark **HANDMADE | E&S MFG CO** has been recovered and may have been made by this company after the name change. Known

brand names that they used were **VICTORY, LUMBERMAN'S PRIDE, DIAMOND, PIONEER,** and **FOREST KING**. See Damon Brothers; also Witherell Scythe Co. Klenman (1990) has several pages devoted to Emerson & Stevens, including a number of photographs of axes with the original company labels on them. Klenman notes that Emerson & Stevens closed in 1965 and that they made hand hammered axes until then, rather than dropped forged all steel axes, as had long been the case with other manufacturers.

Klenman also notes that there were as many as 15 ax factories operating in Oakland on a two mile stretch of Emerson stream, which dropped 110 feet in elevation and provided the water power for these factories. Geoff Burke from the Wooden Boat School attended the Emerson & Steven's 1969 closeout factory sale and was told that the last step in the ax forging process was to cold hammer the axes at room temperature to give them unique qualities of toughness and strength.

Emerson, Luther* Waterville 1880-
Tools Made: Axes
Remarks: See Emerson & Stevens.

Emerton, Joseph B.** Auburn -1855-
Tools Made: Forks
Remarks: He is listed in the 1855 *Maine Registry and Business Directory*.

Emery, C. & W. D.* Sedgwick
Tools Made: Axes

Emery, Cyrus & Co. Sullivan -1869-1883-
Tools Made: Axes, Forks, and Hoes
Remarks: They are listed in the 1869 *Maine Business Directory* as C. Emery & Co.

Cyrus Emery of Sullivan is listed in the 1879 *Maine Business Directory* as an ax-maker. He is listed in the 1881 *Maine Business Directory* as Cyrus Emery & Co., maker of axes, etc.

Emery & Jillson   Kennebunk   -1849-
Tools Made: Edge Tools

Emery, Gilmore   Newfield   -1879-
Tools Made: Plows
Remarks: Possibly he was related to Jeremiah Emery.

Emery, H.   Buxton   -1849-
Tools Made: Edge Tools

Emery, Jeremiah W.   Newfield   -1871-1885-
Tools Made: Farm Tools and Plows, including cultivators
Remarks: In an 1879 *Maine Business Directory*, both he and Gilmore Emery are listed. Their relation, if any, is unknown.

Emery & Waterhouse**   Monmouth   -1882-
Tools Made: Axes
Remarks: They are listed in the 1882 *Maine Business Directory*.

Emery, Waterhouse & Co.   Portland   1846-1885
Tools Made: Wood Planes
Remarks: They were hardware dealers who marked their planes, **EMERY, WATERHOUSE&CO. | PORTLAND.ME.** with the top line curved. Other variations of their name were used before 1871 and after 1894 (Nelson, 1999). Pollak (2001) states that "From 1846 to 1870, the firm logo read **EMERY + WATERHOUSE**. In 1871 it changed to **EMERY WATERHOUSE & CO.** and in 1894 changed once again to **EMERY & WATERHOUSE CO.**"

Enos, Emery**   Montville   -1880-
Tools Made: Blacksmith
Remarks: H is listed in the 1880 census.

Evans, George Franklin   Norway   ca. 1862-1864-   (b.1842, d.1904)
Tools Made: Metal Planes and Gunsmith
Remarks: Evans held two patents (28 Jan. 1862 and 22 March 1864) for iron circular planes. No known specimen of the 1862 patent has been found. However, the 1864 patent led to the 1864-1871 production of iron circular planes. Some of these may have been

made in Bennett's Mills, a foundry in Norway, which he purchased in 1864, but otherwise they were produced by R. H. Mitchell & Co. out of Hudson, NY, Michael Schwartz in Bangor, the Circular Plane Co., and Darling & Schwartz out of Bangor (Nelson, 1999). R. H. Mitchell & Co. purchased patent rights and made most of the Evans circular planes, all of which had brass name plates. The Davistown Museum has an Evans circular plane (ID# TJE1001) that shows evidence of once having a name plate, but one is no longer present, in the IR collection. "It is believed that about the same time Bailey was developing his No. 13 circular plane (1871) the Stanley Rule & Level Company negotiated with the R. H. Mitchell Company and purchased their plane business" (Smith 1981). With his brother Warren, G. F. Evans formed the Evans Rifle Mfg. Co. in 1873. Later, he became superintendent of the Diamond Wrench Mfg. Co. in Portland in 1886. He moved out of state to Somerville, MA in 1888 and made and patented a machine speed regulator.

Evans Rule**
Remarks: No information available

Faber, John   Bangor   -1859-
Tools Made: Edge Tools

Fabian, Andrew   Bangor
Tools Made: Rules, including Log Rules
Remarks: His relationship to Valentine Fabian is unknown.

Fabian, Valentine   Milo Junction   -1897-1930
Tools Made: Rules, including Log Rules
Remarks: Fabian moved to Milo Junction from Bangor in 1897. He was also reported living in Orneville, ME, but the dates are not known. He marked his rules: **V. FABIAN | SOLE MANUFACTURER, MILO JCT.ME | COPYRIGHT SECURED**; however, someone continued to use his name and make rules after his death (Nelson, 1999). The Davistown Museum has a log caliper made by Fabian (ID# 51102T1) in the IR collection. The Fabian log caliper in

these photos is three feet long with 20" arms and is in the collection of Michael Horn of Ogunquit, Maine.

Fairbanks Bros.   Mt. Vernon   -1879-
Tools Made: Rakes

Fairbanks, Charles   Belgrade   -1879-
Tools Made: Rakes

Fairbanks, Joseph   Monmouth   -1855-1856-
Tools Made: Edge Tools
Remarks: He is listed in the 1855 and 1856 *Maine Business Directories.*

Fairbanks, Sylvannus   Mt. Vernon   -1871-
Tools Made: Rakes

Fairfield, Edward**   Portland   1858
Tools Made: Gunsmith
Remarks: Demeritt (1973, 163).

Farnham, S. & Son**   Bucksport   -1856-
Tools Made: Shipsmith
Remarks: They are listed in the 1856 *Maine Business Directory.*

Farnham, William   Richmond   -1856-
Tools Made: Edge Tools
Remarks: He is listed in the 1856 *Maine Business Directory.* The Davistown Museum has a drawknife (ID# 100605T4) marked **W. FARNHAM** in the MIV collection.

Farrar, J.   Buckfield   -1879-
Tools Made: Farm Tools and Shovels (Snow Shovels)

Farrington, George**   Waldoboro   1833-
Tools Made: Blacksmith
Remarks: A native of Warren, in 1833, when a young man, he moved to Waldoboro where he remained the rest of his life. He was a blacksmith, but subsequently gave up his trade to engage in farming (*Biographical Review* 1897, 245).

Felch, Isaac N.**   Belfast   -1856-
Tools Made: Wood Lathes
Remarks: Jeff Joslin indicates that, based on patent information, Milton Roberts and H. E. Pierce were lathe makers working as partners in 1854. By 1856, Roberts had a new partner, Isaac N. Felch. All of their patents related to wood lathes for production use.

Fernald, Donald   Saco   1816-1832
Tools Made: Edge Tools

Fickett, George**   Portland   -1855-
Tools Made: Pump and Block Makers
Remarks: He is listed in the 1855 *Maine Registry and Business Directory.*

Field, David**   Milbridge   -1855-
Tools Made: Pump and Block Makers
*Remark*s: He is listed in the 1855 *Maine Registry and Business Directory.*

Fillebrown, L. W.   Wayne   -1879-1907-
Tools Made: Farm Tools, including Harrows and Cultivators, Lathe Knives
Remarks: He is listed in the 1881 *Maine Business Directory.* See Fuller, R. C. & Co.

Finley, A. R.**   Lincoln   -1867-
Tools Made: Axes
Remarks: He is listed in the 1867 *Maine Business Directory.*

Finney, A. R.   Atkinson, Milo   -1855-1862
Tools Made: Axes and Edge Tools
Remarks: There was a Finney in Atkinson listed in the 1855 and 1856 *Maine Business Directories* and a Finney in Milo listed in 1862 who made iron components for carriages as well as axes. The two are assumed to be the same man.

Fisher & Martin   Newport   -1846-
Tools Made: Vises
Remarks: Comprised of M. Fisher and William Martin Jr., the company made a vise invented by someone reportedly named Matthews. It is not known if M. Fisher could have been the same man as Mark Fisher of Levant.

Fisher, Mark   Levant   1843-1847
Tools Made: Anvils
Remarks: Starting in Maine and moving later to Trenton, New Jersey in 1847, Fisher was a prolific anvil maker. In NJ, he became part of the company Fisher & Norris (Eagle Anvil Works); his anvils continued to be marked: **FISHER MAKER | PATENT APRIL 24 1877**. His son Clark Fisher continued to make anvils after him, changing the company to Eagle Anvil, Vise & Joint Works.

Fitton, S.   Buckfield   -1871-
Tools Made: Rakes

Fletcher & Smith**   West Buxton, Hollis   -1879-
Tools Made: Knives
Remarks: This company is listed in the 1879 *Maine Business Directory*. See Thomas Fletcher of Hollis.

Fletcher, A. H.   Harmony   -1849-
Tools Made: Edge Tools

Fletcher, Thomas**   North Hollis, Hollis   -1867-
Tools Made: Shoe Knives
Remarks: He is listed in the 1867 *Maine Business Directory*.

Flint, Nathan R.   West Baldwin   -1871-
Tools Made: Farm Tools, including Ax Handles

Flowers, Richard**   Hope   -1855-
Tools Made: Edge Tools
Remarks: He is listed in the 1855 *Maine Business Directory*.

Floyd   Portland   ca. 1850
Tools Made: Adzes and Axes
Remarks: His tools are marked **FLOYD | PORTLAND ME**. One tool has been reported marked by both Floyd and E. G. Bolton. The Davistown Museum has a Floyd adz in the MIV collection.

Floyd & Stanwood   Portland   -1855-1856-
Tools Made: Edge Tools
Remarks: This company is listed in the 1855 and 1856 *Maine Business Directories*.

Flyn, John**   Warren   -1780-1800(?)
Tools Made: Planes
Remarks: The Davistown Museum has a complex moulding plane (photos) marked **JOHNFLYN** in the MII collection. This tool was found in Warren, Maine, by the tool collector Ben Blumenberg, and it is believed to have been made in the late-18th century in the vicinity of Warren or Thomaston, Maine. The plane is characterized by heavy chamfering on the sides and unique curves at the heel of the moulding and is made of yellow birch. This complex moulding plane may represent the work of a Maine

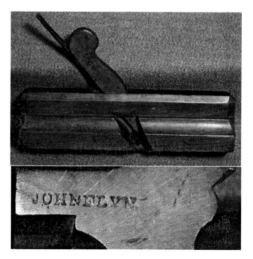

planemaker, who predates the plane production of Joseph Metcalf and Thomas Waterman, until now considered Maine's first documented planemakers.

Fogg, A. M. & Co.   Houlton   -1871-1900-
Tools Made: Farm Tools

Fogg, H.**   -1810-1830-
Tools Made: Planes
Remarks: Pollack (2001, 151) lists a 29 3/4" and a 22 1/2" jointer with shallow, round chamfers, a diamond strike, and three bead planes, all beech, all marked **H. FOGG**. One of the jointers was found in VT, and the beads in NH, one of which is also marked **T.B. Gove** suggesting a northern New England origin. See C. Gove.

Fogg, Jonathan   Bridgton   -1871-
Tools Made: Farm Tools

Folsom, I. J.   East Livermore   -1885-
Tools Made: Plows

Foss, Benjamin F.**   Fairfield
Tools Made: Screwdrivers
Remarks: Foss was issued patent 873363, 12/10/1907 for a ratchet screwdriver, information courtesy of C.D. Fales.

Foster, A. F.   Nobleboro   -1869-
Tools Made: Farm Tools, including Corn Shellers

Foster, Albert W.   Fairfield   -1849-
Tools Made: Edge Tools

Foster, Andrew**   Machias, East Machias   -1850-1870-
Tools Made: Gunsmith and Jeweler
Remarks: Demeritt (1973, 163) states he moved to E. Machias in the 1870s.

Foster, Henry**   Lubec   -1870-
Tools Made: Gunsmith
Remarks: Demeritt (1973, 163).

Foster, Henry M.   Skowhegan, Gardiner   -1865-1882-
Tools Made: Axes
Remarks: Foster moved from Skowhegan to Gardiner around 1881. His name has also been recorded as H. N. Foster. He is listed in Skowhegan in the 1869, 1879, and 1881

Maine Business Directories and in Gardiner at Dam No. 3 in the 1882 *Maine Business Directory*.

Foster, J.**   Bath   -1850-
Tools Made: Planes
Remarks: "This imprint was reported on a jointer of mid-19c. appearance. [Marked] **J. FOSTER BATH ME**" (Pollak 2001, 154). The Bob Jones collection has two planes marked J. Foster. The ovolo and cove moulding plane matches the mark that Pollak believes is from Jesse Foster, a Boston, MA, turner and cabinetmaker. The other, a 9 ½" mitre plane, appears more like the Bath mark but does not give any location.

Foster, Jacob**   East Machias   -1870-
Tools Made: Gunsmith
Remarks: Demeritt (1973, 163).

Foster, Lewis*   Machias   1882-1884
Tools Made: Axes

Fowler, James   Portland   -1862-
Tools Made: Cutlery and Medical Tools

Fox, George H.   Bangor   -1899-1902-
Tools Made: Household Tools
Remarks: Ice cream molds with the patent dates of 14 Feb. 1899 and 15 July 1902 were either made or patented by Fox.

Frederick & Swan & Co.**   -1830-1840-
Tools Made: Planes
Remarks: Pollak (2001, 401) lists an example, marked **FREDERICK & SWAN | &CO | MANUFACTURERS**, of a 11" razee boxwood smoother with a closed handle, a double-iron, and heavy round chamfers, found in Maine. The **FREDERICK & SWAN** is in an arch.

Freeman, G. H. & Co.   Presque Isle   -1899-1900-
Tools Made: Farm Tools

French, A.   Hampden Hlds (?)
Tools Made: Rules
Remarks: Most often using the mark **A. FRENCH | HAMPDEN HLDS, ME** on his log rules, he sometimes added the middle initial "B" or "E".

French, Benjamin**   Brunswick   -1838-
     Tools Made: Blacksmith

French, Frederic**   Warren   -1882-
     Tools Made: Edge Tools
     Remarks: He is listed in the 1882 *Maine Business Directory*.

French, John L.   Chelsea   -1830-1880-   (d.1884)
     Tools Made: Blacksmith, Axes, and Edge Tools
     Remarks: He is listed in the 1867 and 1869 *Maine Business Directories*. Other information is from DATM (Nelson 1999) and Henry D. Kingsbury and Simeon L. Deyo, 1892, *Illustrated History of Kennebec County, Maine*. Blake, NY.

Frost, Benjamin*   Westbrook   1927-
     Tools Made: Axes

Frost, Harrison**   Mariaville   -1855-
     Tools Made: Forks
     Remarks: He is listed in the 1855 *Maine Registry and Business Directory*.

Frost, L. W.**   Bridgton   -1881-
     Tools Made: Axes
     Remarks: He is listed in the 1881 *Maine Business Directory*.

Fry, Isaiah   North Berwick   -1855-
     Tools Made: Farm Tools
     Remarks: See Frye & Son, Isaiah and Frye & Co.

Frye   Portland
     Tools Made: Medical Tools
     Remarks: "Frye was only listed as a maker of 'instruments', but the context implies medical items and also that he worked before 1870" (Nelson, 1999).

Frye & Co.**   Portland
     Tools Made: Plows
     Remarks: According to Rathbone (1999), Frye & Co.

was established by Isaiah Frye, father of John J. Frye. The wrench in the figures was found in a field in S. Portland. It is very similar to the wrench pictured in the book.

Frye, Ensign**   Skowhegan   -1855-
Tools Made: Forks
Remarks: He is listed in the 1855 *Maine Registry and Business Directory.*

Frye, Isaiah & Son   Portland   -1869-1871-
Tools Made: Plows
Remarks: This Isaiah Frye is probably the same as the Isaiah Fry reported in North Berwick.

Frye, John J.   Portland   -1879-1885-
Tools Made: Farm Tools and Plows, including Cultivators and Mowing Machines

Frye, Joseph   Fryeburg
Tools Made: Scientific Instruments
Remarks: Frye was a surveyor when Fryeburg was still a part of Massachusetts.

Fuller, C. & D.
Tools Made: Wood Planes
Remarks: The mark **C. FULLER | D. FULLER** has a similarity to the mark of David Fuller (**D. FULLER**) of West Gardiner, Maine. "The C.Fuller part of the mark does not match any of the known marks of Charles Fuller of Boston, MA, but there are gaps in his pre-1850 history when he could have worked in ME; conversely, David could have worked in MA (he was born there) before moving to ME" (Nelson 1999).

Fuller, David   West Gardiner   -1829-1856-   (b.1795, d.1871)
Tools Made: Wood Planes
Remarks: Although born in Ipswich, MA, Fuller was listed as a planemaker in 1829 and then again in 1855-56 in West Gardiner, ME. In the middle years, he was a carpenter, cabinetmaker, joiner and/or architect, but probably continued to make planes as indicated by personal papers. In the early 1850s, his son, Erastus, was selling his planes out of Bath, ME. His planes were marked either **D. FULLER** or **D. FULLER | GARDINER** (Pollak 2001, 158). The Bob Jones collection has a 9 ½" beech single boxed plane with the Gardiner mark and a 9 ½" Roman ogee plane marked **D.FULLER | E.FULLER | 1851**. They are both mentioned in Pollak (2001). Butterworth and Blumenberg (1991a, 1991b) have written two articles on Fuller in *The Chronicle*. The Davistown Museum has a Fuller rabbet plane in the Maritime IV collection (photo).

Fuller, Leonard**   Farmingdale   -1855-
Tools Made: Edge Tools
Remarks: He is listed in the 1855 *Maine Business Directory*.

Fuller, R. C. & Co.   Wilton   -1885-
Tools Made: Farm Tools
Remarks: The company advertised "Fillebrown harrows," probably using L.W. Fillebrown's (of Wayne, ME) patent for harrow teeth.

Fuller, Silas**   Belfast   -1855-
Tools Made: Pump and Block Makers
Remarks: He is listed in the 1855 *Maine Registry and Business Directory*.

Fundt, John J.**   Hartland   -1867-
Tools Made: Axes
Remarks: He is listed in the 1867 *Maine Business Directory*.

Furbish, John   Brunswick   -1879-
Tools Made: Farm Tools

Furbish, Zachary T.**   Augusta   1870-1896
Tools Made: Screwdrivers
Remarks: Furbish was one of Maine's most famous 19th century inventors and toolmakers and was also a carpenter and machinist.
There is much more information on his work and patents in the museum's Furbish company file in volume 8 of the Hand Tools in History series (Brack 2008b). The screwdriver in the photograph 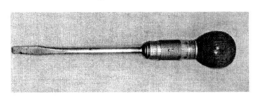 (courtesy of C.D. Fales) is Furbish's **FOREST CITY SCREWDRIVER CO. | PATENTED APRIL 16, 1895.** Subsequent Furbish patents were assigned by him to North Brothers Manufacturing Co. of Philadelphia, PA.

Furbush, Charles**   Belfast   -1881-
Tools Made: Shipsmith
Remarks: He is listed in the 1881 *Maine Business Directory*.

Gaffield, Isaac   Augusta   -1855-
Tools Made: Farm Tools

Gale, Gin & Co.   Wayne   -1839-
Tools Made: Scythes

Gammage**   ca. 1820

Tools Made: Planes

Remarks: Pollak (2001, 161) lists this mark with no location. The Bob Jones collection has one plane with this mark that was found in Maine. Bob St. Peters has a 10 3/8" long hollow plane made of birch marked **T. Gam___e** (possibly Gammage) that is clearly American-made.

Gammon, E. & Co.   Gorham   -1849-

Tools Made: Edge Tools

Gammon, W. W.   Dixfield   -1885-

Tools Made: Rakes

Garcelon, Charles E., Jr.**   Lewiston   1893

Tools Made: Gunsmith

Remarks: Demeritt (1973, 163).

Gardner Edge Tool Co.*   Gardner   1889-1899

Tools Made: Axes

Gardner, C. S.**   East Machias   -1856-

Tools Made: Shipsmith

Remarks: He is listed in the 1856 *Maine Business Directory*.

Garity, J.**   -1850-

Tools Made: Planes

Remarks: Pollak (2001, 162) lists 4 rosewood and 2 lignum vitae bench planes, all from the Maine coast area. (Note: there is no signature on the planes.)

Garland, Einanan   Kenduskeag   -1855-

Tools Made: Farm Tools

Garland Manufacturing Company**   Saco   1866 - present day

Tools Made: Mallets and Hammers

Remarks: The company was founded in 1866 and was incorporated in 1873. For more information, see the museum's company file in vol. 8 (Brack 2008b) on Garland.

Gay & Parsons   Augusta   -1888-

Tools Made: Screwdrivers, specifically ratchet screwdrivers, and Gunsmith

Remarks: Demeritt (1973, 164) indicates that George Gay was also a gunsmith when working with this company. See Gay, George E.

Gay, George E. Augusta -1878-1905-
Tools Made: Screwdrivers
Remarks: Gay had three patents for different types of screwdrivers - a ratchet screwdriver (patented 1878), a spiral screwdriver (patented 1892), and another ratchet screwdriver (patented 29 July 1902). Except when he was part of Gay & Parsons for a few years around 1888, George worked alone. He used several marks to distinguish his screwdrivers: **GAY | PAT. DEC 17, 1878 || GEO.E.GAY | AUGUSTA | ME.** and **GEO.E.GAY AUGUSTA, MAINE U.S.A. | PAT. DEC. 13, 1892.** It is thought that he added "Mfg. Co." to his name and acquired the brand name **UNION** after 1900. According to C.D. Fales (personal communications), he shared with John A. Parsons patent 210942, 12/17/1878, for an improvement in ratchet screwdrivers; patent 437297, 9/30/1890, ratchet screwdriver; patent 484004, 10/11/1892, spiral screwdriver; and, by himself, patent 644907, 3/6/1900, screwdriver and patent 705917, 7/29/1902, patent screwdrivers. The Davistown Museum has a Gay screwdriver with the December 17th patent date in the IR collection (ID# 32708T59).

Gem Auburn -1865-1895-
Tools Made: Household Tools
Remarks: This name has been found on three different tools. The first, a flour sifter, bears the patent of J. Wells, Brooklyn, NY, 26 Dec. 1865, the second is an apple corer patented by James Fallows, Philadelphia, PA, 2 Jan. 1877, and the third is a raisin seeder bearing an 1895 patent and the Auburn, ME, location. The DATM (Nelson 1999) suggests that three different makers were using the "Gem" name.

George, H. M.** Lewiston -1881-
Tools Made: Edge Tools
Remarks: He is listed at 26 Bates St. in the 1881 *Maine Business Directory.*

Gerry, Wm. P.** Robbinston -1856-
Tools Made: Shipsmith
Remarks: He is listed in the 1856 *Maine Business Directory.*

Getchell, John S. & Sons Houlton -1879-1900-
Tools Made: Farm Tools, Hoes, and Plows, including Harrows.

Getchell, James M.** Bath 1867-1872
Tools Made: Gunsmith
Remarks: Demeritt (1973, 164).

Gibson, Kimball & Sanford Bangor -1869-1871
Tools Made: Saws

Remarks: The company began as the Dirigo Saw Works and was succeeded in 1871 or 1872 by Kimball & Sanford. Sanford's name has also appeared spelled as Sandford.

Gilbert, Daniel (see Asa Jones)

Gilman, Ellis A.**   Liberty   -1860-1870-
Tools Made: Blacksmith
Remarks: He is listed in the 1860 and 1870 census.

Gilman, Hollis M.**   Liberty   -1850-
Tools Made: Blacksmith
Remarks: He is listed in the 1850 census.

Gilman, John N.**   Liberty   -1880-
Tools Made: Machinist
Remarks: He is listed in the 1880 census.

Gilman, Moses   South Sangerville   -1871-
Tools Made: Farm Tools, including Winnowing Machines

Gilman, Richard H.   Liberty   -1860-1885-
Tools Made: Blacksmith, Farm Tools, Household Tools, and Plows, including Harrows and Cider Presses
Remarks: He is listed in the 1860, 1870, and 1880 census.

Gilman, Roscoe**   Liberty   -1860-
Tools Made: Blacksmith
Remarks: He is listed in the 1860 census living in the household of his father, Richard Gilman.

Gilmore, John Sr.**   Searsport   1784-
Tools Made: Blacksmith
Remarks: He came to Searsport in 1784 and built a temporary log home. He was a blacksmith by trade and manufactured the brick out of which he built one of only two brick houses on the Belfast Road (*Biographical Review* 1897, 284).

Glenn, James H.   Caribou   -1899-1900-
Tools Made: Farm Tools

Glidden & Sons**   Whitefield   -1867-
Tools Made: Edge Tools
Remarks: This company is listed in the 1867 *Maine Business Directory*.

Glidden, Hiram   Waldoboro   -1856-
Tools Made: Edge Tools
Remarks: He is listed in the 1856 *Maine Business Directory*.

Glidden, Joseph   Liberty   -1879-1880-
Tools Made: Farm Tools, including Cultivators

Glover*   Deer Isle   -1874-
Tools Made: Axes
Remarks: Tim Bonelli provided this photo of an ax marked
I.A. GLOVER.

Goddard, Silas   Brunswick   -1867-1885-
Tools Made: Plows

Godfrey, Otis S.**   Cherryfield   -1856-
Tools Made: Shipsmith
Remarks: He is listed in the 1856 *Maine Business Directory*.

Goodhue, I. W. & G. W.**   Bangor   -1855-
Tools Made: Whips
Remarks: They are listed in the 1855 *Maine Registry and Business Directory*.

Goodhue, J. G.(?)**   -1810-1820-
Tools Made: Planes
Remarks: Pollak (2001, 173) lists a 7" lignum vitae smoothing plane marked **J. W. GOODHUE | BANGOR. ME.** made by P.B. Rider/Bangor.

Goodhue, J. W.**   Fort Fairfield   1879-1892
Tools Made: Gunsmith
Remarks: Demeritt (1973, 164).

Goodnow, J. C.   Farmington   -1867-1871-
Tools Made: Axes and Edge Tools
Remarks: He is listed as an edge toolmaker in the 1867 and 1869 *Maine Business Directories*.

Goodrich, Samuel**   Newfield   -1855-
Tools Made: Edge Tools
Remarks: He is listed in the 1855 *Maine Business Directory*.

Goodwin, L. B.**   Surry   1871

Tools Made: Gunsmith
Remarks: Demeritt (1973, 164).

Googins, J. F. & Co.　Biddeford　-1899-1900-
Tools Made: Farm Tools

Gordon & Son　Garland　-1869-
Tools Made: Rakes, specifically Horse Rakes

Goth, Ferdinand**　Portland　1882-1885
Tools Made: Gunsmith
Remarks: Demeritt (1973, 164) states he worked at 40 Market St.

Goth, Frederic**　Biddeford, Portland　1859-1887
Tools Made: Gunsmith
Remarks: Demeritt (1973, 86-8, 164) notes he moved to Portland (40 Market St.) in 1860.

Goth, Richard**　Portland　1882-1896
Tools Made: Gunsmith
Remarks: Demeritt (1973, 164) states he worked at 40 Market St.

Goth, William**　Portland, Augusta　1883-1896-
Tools Made: Gunsmith
Remarks: Demeritt (1973, 164) states he worked at 40 Market St. until 1896, when he moved to Augusta.

Gould, S. L. & Co.　Skowhegan　-1869-
Tools Made: Farm Tools

Gould, Nehemiah**　Dixmont　1875
Tools Made: Gunsmith
Remarks: Demeritt (1973, 164).

Gould, O.　Portland
Tools Made: Axes

Gould, S. S.　Anson　-1879-1880-
Tools Made: Farm Tools

Goulding, William**　Portland　1827

Tools Made: Gunsmith
Remarks: Demeritt (1973, 30, 164).

Gove, C.** Kittery ca. 1840
Tools Made: Planes
Remarks: The Davistown Museum has a complex plow plane (ID# 71504T1) marked **C·GOVE** in the MII collection (photo). This tool was found in Eliot, Maine. It is a typical 18th century style New England plow plane with a unique rosewood fence and adjustment screw. The signature is in 18th century style script. This is the very same plane noted in the 4th edition of Pollak (2001) with the following entry "Possibly Charles C. Gove (d. 1893) a carpenter whose shop burned in 1840, set on fire by varnish boiling over upon a hot stove. Example: a 9" beech plow with one rosewood thumbscrew, rosewood fence, flat chamfers, found in Kittery, ME. ca. 1840-50. UR." Pollak lists 5 other Goves as planemakers, all of whom appear to be late 18th century or early 19th century Gulf of Maine planemakers working within a 75-mile range of the New England coastline, raising the possibility of a family of planemakers by the name of Gove. The museum seeks additional information about "C.Gove" and his working dates. There is a probability that C. Gove's working dates are earlier

than those of Joseph Metcalf or Thomas Waterman, currently considered Maine's first documented planemakers. Also see the listing for the John Flyn moulding plane.

Graham, Edmund H.** Biddeford 1854-1880-
Tools Made: Gunsmith
Remarks: Demeritt (1973, 93-8, 164) notes he later moved to Manchester, NH.

Grant, Stephen Sr.** Berwick, Ellsworth, Bucksport, Monroe ca. 1800
Tools Made: Blacksmith
Remarks: He was born in Berwick. He served as a blacksmith apprentice for seven years, moved to Ellsworth and worked as a blacksmith, then moved to Bucksport working as a carpenter and did the first iron work on a vessel sent out from that port. In 1811 he visited Monroe and purchased a tract of wild land from Thorndike, Prescott, and

Sears. In 1813 he moved onto his land, erected a log cabin, and later a frame house. He was the only blacksmith in the neighborhood (*Biographical Review* 1897, 190-1).

Grant, N.** Yarmouth -1855-
Tools Made: Pump and Block Makers
Remarks: He is listed in the 1855 *Maine Registry and Business Directories*.

Graves & Long** Bangor 1846-1876
Tools Made: Gunsmith
Remarks: Demeritt (1973, 77-9, 164) indicates this is Joseph Graves and Malcolm W. Long.

Graves, B. Solon 1840-1850-
Tools Made: Axes
Remarks: A large hewing ax (ID# 21201T1), probably used for hewing sills or possibly keels, was donated to the Davistown Museum by Rick Floyd of Newport, Maine, and is in the MIII collection (photo). Other than the maker's mark **B GRAVES** and its location **SOLON**, the ax is otherwise unmarked. There is no listing for Graves either in DATM (Nelson 1999) or in Yeaton's (2000) *Axe Makers of Maine*.

Graves, Elisha Calais -1865-1871-
Tools Made: Adzes, Axes, Chisels, and Edge Tools
Remarks: He is listed in Milltown, NB, in 1865, and in Calais, ME, from 1869-71.

Graves, H.** Springfield -1881-1882-
Tools Made: Axes
Remarks: He is listed in the 1881 and 1882 *Maine Business Directories*.

Graves, Robert S.** Bangor 1864-1868
Tools Made: Gunsmith
Remarks: Demeritt (1973, 164).

Gray, Arthur Naples -1867-
Tools Made: Wood Planes
Remarks: Arthur Gray of Naples, ME, was issued Patent No. 65,562 on June 11, 1867 for a transitional jack plane with a heavy cast iron frog attached behind the throat with four wood screws. **PAT. JUNE 11, 1867** was cast into the lever cap (Smith 1991, 86; Pollak 2001, 175).

Greely, John   Palermo   -1869-
Tools Made: Rakes

Green, S.   Winthrop   -1835-
Tools Made: Leather Tools
Remarks: He used the mark **S. GREEN/CAST STEEL** on his tools; he also advertised "shoe tools" at this time.

Greenleaf, Charles T.   Bath   -1867-
Tools Made: Farm Tools

Greenwood, Aaron   South Gardiner
Tools Made: Rules, including Log Calipers

Grey, J. B.   Fort Fairfield   -1899-1900-
Tools Made: Farm Tools?

Grimes, E. P.   Caribou   -1899-1900-
Tools Made: Farm Tools

Griswold, Virgil**   Portland   -1855-
Tools Made: Whips
Remarks: He is listed in the 1855 *Maine Registry and Business Directory*.

Grover, C. W.**   Bremen   1877
Tools Made: Gunsmith
Remarks: Demeritt (1973, 164).

Grover, Simon**   Skowhegan   1896-1898
Tools Made: Gunsmith
Remarks: Demeritt (1973, 164).

Gross & Owen**   Brunswick   -1845-
Tools Made: Blacksmith

Gross, A. K. P.**   Orland   -1856-
Tools Made: Shipsmith
Remarks: He is listed in the 1856 *Maine Business Directory*.

Grueby, George H.**   Portland   1844-1850-
Tools Made: Gunsmith and Jeweler
Remarks: Demeritt (1973, 164).

Guptill, R. P.** Harrington -1870-
Tools Made: Gunsmith
Remarks: Demeritt (1973, 164).

H. & B. Mfg. Co. West Waterville
Tools Made: Hammers and Hatchets
Remarks: Thought to be Hubbard & Blake Mfg. Co., they marked their tools with
H.&B.MFG.CO. | WEST WATERVILLE, ME.

H., D. (see D. H.)

Hadley, William W.** Moluncus -1855-
Tools Made: Edge Tools
Remarks: He is listed in the 1855 *Maine Business Directory*.

Hahn, S. B.** Thomaston -1881-1882-
Tools Made: Shipsmith
Remarks: He is listed in the 1881 and 1882 *Maine Business Directories*.

Haines & Smith** Portland -1869-
Tools Made: Planes
Remarks: A 28" jointer plane with a closed tote marked **HAINES & SMITH |
PORTLAND** is in the Bob Jones collection. Pollak (2001, 181) states Haines & Smith was
a hardware dealer listed in Portland, ME, active in 1869. Example: a beech joiner, ca.
1850. Also see Haynes & Smith.

Hale & Jordan** West Waterville -1855-
Tools Made: Agricultural Implements
Remarks: They are listed in the 1855 *Maine Registry and Business Directory*.

Hale & Stevens Waterville -1849-1857-
Tools Made: Hoes and Scythes
Remarks: See Hale, Samuel & Eusebius.

Hale, Samuel & Eusebius Waterville 1839-1845
Tools Made: Scythes
Remarks: Either one or both of them were part of Larned & Hale (1836 - 1839), and
one may have been involved with Hale & Stevens.

Hall* Norway 1832-1846
Tools Made: Axes

Hall, Albert   Bristol   -1856-
Tools Made: Axes

Hall, Edward**   Bristol   -1855-
Tools Made: Edge Tools
Remarks: He is listed in the 1855 *Maine Business Directory* as an edge toolmaker.

Hall, G. A.**   Abbott Village   1877
Tools Made: Gunsmith
Remarks: Demeritt (1973, 164).

Hall, G. A.   Houlton   -1899-1900-
Tools Made: Farm Tools

Hall, George**   Kennebunkport   -1869-
Tools Made: Shipsmith
Remarks: He is listed in the 1869 *Maine Business Directory*.

Hall, J.**   ?   ca. 1790-1800
Tools Made: Planes
Remarks: Pollak (2001, 182) mentions four planes with the mark **J HALL**. These include the 9 ½" plow plane with flat chamfers that is in the Bob Jones collection. Jones notes that the plane is in the 18th century style. It is possible that J. Hall is one of the few known 18th century Maine planemakers.

Hall, J. S.*   Portland
Tools Made: Axes
Remarks: It is not known if this is a different J. S. Hall then the one in Fort Fairfield because his working dates are not known.

Hall, J. S.   Fort Fairfield   -1867-1871-
Tools Made: Axes
Remarks: He is listed in the 1867 and 1869 *Maine Business Directories*.

Hall, John Hancock**   Portland, North Yarmouth   -1810-1818   (d. 1841)
Tools Made: Machinist Tools, Rifles, and Gunsmith
Remarks: Hall was an important Maine gunsmith who designed, patented, and invented a breech loading rifle with interchangeable parts and made improvements to Simeon North's Middletown, CT, milling machine, which was used in the gun-making business. He went to Harper's Ferry, W. Virginia in 1818 (Demeritt 1973, 13-28, 164).

Hall, James & Son   Windham   -1855-1871-
Tools Made: Farm Tools

Hall, R. C.   Waterville   -1862-
Tools Made: Axes

Hall, Tilden   Damariscotta   -1867-1871-
Tools Made: Axes and Edge Tools
Remarks: He is listed as making axes in the 1867 and 1869 *Maine Business Directories.*

Hall, William   Norway   -1871-
Tools Made: Farm Tools, specifically Cultivators

Hall, William H.**   Portland   1852
Tools Made: Gunsmith
Remarks: Demeritt (1973, 164).

Hallowell Iron Foundry   Hallowell   -1879-1880-
Tools Made: Plows
Remarks: Also used the name George Fuller's Sons.

Ham, J. B.**   Richmond   -1855-
Tools Made: Edge Tools
Remarks: He is listed in the 1855 *Maine Business Directory.*

Hamblen, J. D.   Deering   -1879-1880-
Tools Made: Farm Tools, including Mowing Machines
Remarks: A G. D. Hamlen appears in an 1885 directory at a different address in Deering. It may be the same man or a son if one of the spellings of the name is incorrect.

Hamblin, G. D.**   Saco   -1855-
Tools Made: Pump and Block Makers
Remarks: He is listed in the 1855 *Maine Registry and Business Directory.*

Hamlen, Franklin L.**   Augusta   1884-1885
Tools Made: Screwdrivers
Remarks: See the Zachery T. Furbish biography in vol. 8 of the Hand Tools in History Series (Brack 2008b) for patent information.

Handy & Wetherell   South Norridgewock   -1849-

Tools Made: Hammers
Remarks: See Handy, S.M. and Witherell, Samuel B.

Handy, S. M.   Norridgewock   -1862-
Tools Made: Edge Tools and Farm Tools
Remarks: A blacksmith who made harrows, Handy is thought to have also made froes.

Hanley, George**   Thomaston   -1881-1882-
Tools Made: Shipsmith
Remarks: He is listed in the 1881 and 1882 *Maine Business Directories* on Water St.

Hanover, Thomas   Bangor   -1869-1871-
Tools Made: Edge Tools
Remarks: He is listed in the 1869 *Maine Business Directory*.

Hanscom, R. B.**   Lewiston   1896
Tools Made: Gunsmith and Machinist
Remarks: Demeritt (1973, 164).

Hapgood, G. E.   Anson   -1880-
Tools Made: Farm Tools
Remarks: The name might have been Hadgood rather than Hapgood.

Harding, Daniel C.   West Baldwin   -1871-1879-
Tools Made: Hay Rakes

Harding, Joseph, Jr.**   Baldwin   -1855-
Tools Made: Rakes
Remarks: He is listed in the 1855 *Maine Registry and Business Directory*.

Harding, Orin**   Waterville
Tools Made: Planes
Remarks: Orin Harding is a late 19th century planemaker. The Liberty Tool Co. recovered a number of Harding planes in the winter of 2007, one of which has been donated to the Davistown Museum. The Harding tool chest also included 8 planes made by J. C. Jewett of Waterville, one of which, a sash plane, was donated to the museum.

Hardison, G. R.**   Gouldsboro   -1891-
Tools Made: Buck Saws
Remarks: He had patent 459399 on 9/15/1891 for a buck saw. Source: Graham Stubbs.

Hardy & Sherman*   Belfast   1886-1896
Tools Made: Axes and Chisels
Remarks: It is unknown if George Hardy was one of the partners. James Hill of Warren, Maine, has found a 12" long 1 ¼" socket chisel signed **Hardy & Sherman Belfast.**

Hardy, George*   Belfast   1879-1886
Tools Made: Axes
Remarks: He is listed in the 1879 and 1881 *Maine Business Directories.*

Hare & Prescott**   Newport   -1855-
Tools Made: Agricultural Implements
Remarks: They are listed in the 1855 *Maine Registry and Business Directory.*

Harlow, J.**   Robbinston   -1855-
Tools Made: Wheelwright
Remarks: He is listed in the 1855 *Maine Registry and Business Directory.*

Harmer, D.**   Caribou   -1879-
Tools Made: Edge Tools
Remarks: He is listed in the 1879 *Maine Business Directory.*

Harmon, A.**   Scarborough   -1810-1830-
Tools Made: Planes
Remarks: Pollak (2001, 186) states that A. Harmon is believed to be Abner Harmon, who was listed in the 1810 census for Scarborough, Cumberland Co., ME, adjacent to Buxton. In the 1820-1830 censuses, Abner Harmon was listed in Buxton, York Co., ME. Examples marked **A. HARMON** are a 12 7/8" beech double-blade fixed sash with offset tote and a 8 ½" beech single-boxed center bead.

Harmon, Benjamin   Biddeford, Scarborough, Buxton
Tools Made: Wood Planes
Remarks: Planes marked **B. HARMON | BIDDEFORD** are assumed to be from Maine. Pollak (2001, 186) states that the mark **B.HARMON*** is believed to be from Benjamin Harmon, listed in the 1810 census for Scarborough, adjacent to Buxton. In the 1820-30 censuses, he is listed in Buxton. The Bob Jones collection has a 9 5/8" skewed rabbet plane with flat chamfers and no location on the mark.

Harmon, Horace**   Lubec   -1856-
Tools Made: Shipsmith
Remarks: He is listed in the 1856 *Maine Business Directory.*

Harmon, I., Jr. **   Buxton?   (b. 1791)
Tools Made: Planes
Remarks: Pollak (2001, 187) states that this is possibly Ismiel Harmon (b. 1791), who was in the 1820-1830 census from Buxton, ME and listed as a carpenter in Benton, ME, from 1820-1850 or John Harmon Jr., also listed in the 1820 census from Buxton, York Co., ME. Examples marked **I + HARMON Ir.** (Jr.) are a 10" beech fixed sash with slightly rounded chamfers and a 10 ½" beech complex molder with round chamfers.

Harmon, P.**   Biddeford
Tools Made: Planes
Remarks: Pollak (2001, 187) lists the mark **P.HARMON | BIDDEFORD** as made in Biddeford, Maine. The Bob Jones collection has a 14" razee skewed rabbet plane with this mark that was found in Maine. It is unknown if there is any relationship between P. Harmon and the more common B. Harmon Biddeford mark.

Harper, E. H.   Biddeford   -1869-1871-
Tools Made: Files

Harper, Samuel*   Limerick   -1874-
Tools Made: Axes

Harper, Samuel, Jr.   Waterboro   -1856-
Tools Made: Edge Tools
Remarks: He is listed in the 1856 *Maine Business Directory*.

Harriman, G. M.**   S. Thomaston   -1893-
Tools Made: Saws
Remarks: He had patent 506111 on 10/3/1893 for a saw frame. Source: Graham Stubbs.

Harrington, Francis**   Rockland   -1855-
Tools Made: Pump and Block Makers
Remarks: He is listed in the 1855 *Maine Registry and Business Directory*.

Harris, D. R.**   Brownville   -1877-1882-
Tools Made: Axes, Knives, and Gunsmith
Remarks: He is listed as a gunsmith by Demeritt (1973, 164) from 1877-1878. He is listed in the 1881 and 1882 *Maine Business Directories* as an ax- and knife-maker.

Harris, R.*   Brownville
Tools Made: Axes

114

Harris, T. W.**   Portland   ca. 1850
Tools Made: Planes
Remarks: Pollak (2001, 188) reports an adjustable sash plane with the imprint
T.W.HARRIS. | PORTLAND. The Bob Jones collection has an 11 ½" birch grooving plane
with round chamfers.

Hart, A. W.*   Lincoln Plt. (?)   1902-1911
Tools Made: Axes

Hartley, V. A.   Woodstock   -1862-
Tools Made: Blacksmith and Edge Tools

Harvey, H. H. & Co.   Augusta   -1872-1914-
Tools Made: Blacksmith Tools, Ice Tools, and Stone-working Tools
Remarks: The company manufactured the tools in Maine and sold them out of a
Boston, MA office. The "& Co." part of their name was used inconsistently. Tools were
marked **H. HARVEY | AUGUSTA.ME. | MANUF'R.**

Harvey, William & Sons*   Augusta, Oakland, Readfield   1872-1914-
Tools Made: Axes and Scythes

Remarks: A History of The North Wayne Tool Co.
(Kallop 2003) has been published by the Wayne
Historical Society. It gives some information on William
Harvey & Sons and states on page 95: "Harvey at the
same time is identified with the North Wayne Tool
Company... as shareholder... [and] as one of the three company directors." In the
ephemera collection of Raymond Strout is the pictured receipt, dated June 12, 1893,
Readfield. Lamond (2008b) writes that William Harvey was an entrepreneur who not
only made axes but he was also involved in paper, woolens, lumber, harnesses, and salt.
He was holder or partial holder of five patents for a paper placer, nail extractor, scythe,
ax, and halter. His ax labels included **"WM. HARVEY'S CHARCOAL AXE"**, **"KING-
HARVEY HAND-MADE AXE"**, and **"THE DIRIGO"**.

Haskell & Webb*   Bangor
Tools Made: Axes

Haskell Brothers*   Bangor   1891-1896
Tools Made: Axes

Haskell, George B. Co.   Oakland
 Tools Made: Axes
 Remarks: The brand name **KING** was used.

Haskell, G. G.   Bangor   -1884-1891-
 Tools Made: Axes, Carpenter Tools, Cooper Tools, and Shaves
 Remarks: He marked his tools: **G.G. HASKELL | BANGOR**.

Haskell, Isaac W.   Garland   -1862-
 Tools Made: Blacksmith and Edge Tools
 Remarks: The museum has a drawshave
(ID# 71903T5) marked **I. HASKELL** in the
IR collection (photos).

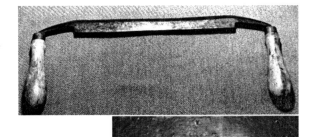

Haskell, Josiah*   Lincoln   1879-1883
 Tools Made: Axes
 Remarks: He is listed in the 1879 and 1882 *Maine
Business Directories*.

Haskell, William**   Carroll   -1867-
 Tools Made: Edge Tools
 Remarks: He is listed in the 1867 *Maine Business Directory*. The museum would
welcome more information on the Haskell clan's edge tool production.

Hatch & Mead**   Castine   -1856-
 Tools Made: Shipsmith
 Remarks: They are listed in the 1856 *Maine Business Directory*.

Hatch, Amos   Jackson   -1879-
 Tools Made: Farm Tools (Mowing Machines)

Hatch, Augustus**   Damariscotta   -1869-1882-
 Tools Made: Shipsmith
 Remarks: He is listed in the 1869, 1881, and 1882 *Maine Business Directories*.

Hatch, Elisha**   Bristol   -1745-
 Tools Made: Blacksmith
 Remarks: He came from Massachusetts and was an early settler in Bristol
(Biographical Review 1897, 106).

Hatch, H. E.   Dexter   -1884-
Tools Made: Clamps
Remarks: May have made or patented a floor clamp.

Hatch, Howland**   Bristol, Bristol Mills   -1775-
Tools Made: Blacksmith
Remarks: Howland was the son of Elisha Hatch and learned the trade from his father (Biographical Review 1897, 106).

Hatch, J. S.**   Waldoboro   -1856-
Tools Made: Shipsmith
Remarks: he is listed in the 1856 *Maine Business Directory*.

Hatch, Jonathan**   Montville   -1860-
Tools Made: Blacksmith
Remarks: He is listed in the 1860 census.

Hatch, S. J.**   Waldoborough   -1869-
Tools Made: Shipsmith
Remarks: He is listed in the 1869 *Maine Business Directory*.

Hathorn, Benj.**   Pittsfield   -1855-
Tools Made: Edge Tools
Remarks: He is listed in the 1855 *Maine Business Directory*.

Haviland, Frederick P.**   Waterville   1838
Tools Made: Gunsmith and Machinist
Remarks: Demeritt (1973, 34-6, 164).

Havner & Rogers**   Searsport   -1869-
Tools Made: Shipsmith
Remarks: They are listed in the 1869 *Maine Business Directory*.

Hawes, C. A.**   Poland   1877
Tools Made: Gunsmith
Remarks: Demeritt (1973, 164).

Hawes, Charles E.   Smithfield   -1880-
Tools Made: Farm Tools, including Barrows and Cultivators

Hawkes, N.   Appleton   -1879-
Tools Made: Farm Tools (Cultivators)

Hawkes & Dresser**   Portland   -1855-
Tools Made: Agricultural Implements
Remarks: They are listed in the 1855 *Maine Registry and Business Directory*.

Hayden, A. R.**   Skowhegan   1891-1897
Tools Made: Gunsmith
Remarks: Demeritt (1973, 164).

Hayden, C. W.**   Skowhegan   1891-1897
Tools Made: Gunsmith and Locksmith
Remarks: Demeritt (1973, 164).

Hayford, A. B.   Millbridge   -1869-
Tools Made: Edge Tools and Shaves
Remarks: His first name may have been Americus. He is listed in the 1869 *Maine Business Directory*.

Hayford, G. S.   Lincoln   -1870-1871-
Tools Made: Edge Tools

Haynes & Pillsbury   Bangor
Tools Made: Saws
Remarks: Their mark was **HAYNES & PILLSBURY | BANGOR.Me | CAST STEEL WARRANTED** with the name line curved and the town and state in a contoured outline.

Haynes & Smith   Portland   -1869-
Tools Made: Wood Planes
Remarks: The DATM (Nelson 1999) indicates that they were hardware dealers who marked planes **HAYNES&SMITH | PORTLAND**; however, Pollak (2001, 185) reports a 9 ½" beech smoothing plane with the imprint **Haines & Smith | Portland**.

Haynes, J.**   Hollis   -1859-1861-
Tools Made: Handsaws
Remarks: J. Haynes had two patents for handsaws: 25015 on 8/9/1859 and 31054 on 1/1/1861. The second has T. T. Lewis of Boston, MA as an assignee. Source: Graham Stubbs.

Heal, Benjamin   Islesboro   -1856-1870-
Tools Made: Axes and Gunsmith
Remarks: Demeritt (1973, 164) notes him as a gunsmith in the 1860s and 1870s.

Heath, W.**   Bath   1876
Tools Made: Gunsmith
Remarks: Demeritt (1973, 164).

Herrick, Eugene I.**   Rangeley   1890
Tools Made: Gunsmith
Remarks: Demeritt (1973, 164).

Hersey, T. & Co.   Paris   -1855-
Tools Made: Farm Tools

Hersey, J. L. B.   Portland   -1834-1849-
Tools Made: Wood Planes
Remarks: Joel Hersey appeared in Portland directories as a joiner in the years of 1834-49 and marked his 9 ½" beech molding planes **J L HERSEY | PORTLAND** (Pollak 2001, 199).

Hersey, S. C.**   ?
Tools Made: Planes
Remarks: Pollak reports three planes with this mark; two were found in Maine (2001, 199). The Bob Jones collection also has a 9 7/8" birch hollow plane with flat chamfers with this mark that was found in Maine.

Hersey, S. S.   Farmington   -1861-
Tools Made: Household Tools
Remarks: An apple parer was patented by Hersey on 18 June 1861; it is unknown if he was also its maker.

Hewlett, Avery   Bowdoinham   -1869-1871-
Tools Made: Axes and Edge Tools
Remarks: The spelling Hewet has also been recorded for Avery's last name. He is listed in the 1869 *Maine Business Directory*.

Higgins*   Bangor   1905-1906
Tools Made: Axes

Higgins & Libby   Portland   -1855-1856-
Tools Made: Axes and Chisels
Remarks: Marked tools: **HIGGINS & LIBBY | PORTLAND**. This company is listed in the 1855 and 1856 *Maine Business Directories*. See

Libby & Bolton. The Museum has a Higgins & Libby gouge (ID# 61204T17, photo) and drawshave (ID# 61204T3) in the MIV collection.

Higgins & Webb   Bangor   -1867-1869-
Tools Made: Axes and Edge Tools
Remarks: Their tools are marked **HIGGINS & | L.WEBB**. See Higgins, J. & A. M. and Webb, Lester. Higgins & Webb are listed in the 1867 and 1869 *Maine Business Directories*.

Higgins, J. & A. M.   Bangor   -1849-
Tools Made: Axes, Drawknives and Edge Tools
Remarks: The mark **HIGGINS ‖ BANGOR** was found on a drawknife and an ax. These edge toolmakers were listed in the 1849 *Maine Business Directory*. One or both of these Higgins were most likely involved with Higgins & Webb at a later date.

Higgins, Jefferson* *   Bangor   -1855-
Tools Made: Edge Tools
Remarks: He is listed in the 1855 *Maine Business Directory* as an edge toolmaker.

Higgins, William   Kenduskeag   -1879-
Tools Made: Farm Tools and Rakes, including Horse Rakes and Cultivators

Hight, Amos   Scarboro   1832-1856-
Tools Made: Axes and Edge Tools
Remarks: He is listed in the 1855 and 1856 *Maine Business Directories*. The museum has a distinctly hand-forged hewing ax (photo) with the mark **A HIGHT SCARBORO** (ID# 12801T5) in the MIII collection.

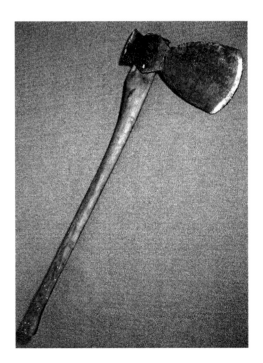

Hight, George   Gorham   1815-1856-
Tools Made: Axes, Edge Tools, and Leather Tools, including Currier's Knives
Remarks: He is listed in the 1855 and 1856 *Maine Business Directories* as an edge toolmaker.

Hilbrook* *
Tools Made: Axes
Remarks: A Kent pattern hewing ax marked

HILBROOK was found in the Hallowell area of Maine and is now in the collection of Donald Bayrd of Milbridge, Maine (photo). This maker is not listed in DATM (Nelson 1999). More information is welcomed.

Hill, Em.** Moscow 1880-1881
Tools Made: Smith
Remarks: Hoyt (1881) lists Em. Hill. Note there also is a Hunnewell & Hill in Moscow in 1879; perhaps this is the same Hill.

Hill, Jonathan** Falmouth (now Portland) 1689
Tools Made: Gunsmith
Remarks: Demeritt (1973, 164).

Hillman & Randall** Fryeburg -1855-
Tools Made: Agricultural Implements
Remarks: They are listed in the 1855 *Maine Registry and Business Directory*.

Hills, Samuel** Union -1786- (b. 1760 d. 1829)
Tools Made: Blacksmith
Remarks: Samuel was the first blacksmith to take up residence in the settlement of Sterlington (Union). He was deaf. Born in Pawtucket, RI, in 1760, he died of consumption in 1829 (Sibley 1851, 58).

Hilton, J.** Kennebunk Landing -1810-1830-
Tools Made: Planes
Remarks: Pollak states that J. Hilton is believed to be John H. Hilton, a cabinetmaker in Kennebunk Landing, ME, active in 1823. Examples marked **J. HILTON** include a 21 ½" shoot board plane in beech, designed as an oversized jack rabbet turned on its side, a 24" joiner, an unusual 10 ¼" beech split sash with screw arms, rosewood nuts with mother-of-pearl inset rings centered with silver buttons and shallow round chamfers, a 9 ½" beech sash coping plane with a tote on the side and flat chamfers, and a level. The sash and coping planes were found on the coast of Maine (Pollak 2001, 203).

Hilton, William O.* Greenville 1893-1908
Tools Made: Axes

Hinckley and Egery**   Bangor   -1855-1871-
Tools Made: Saw Irons, Agricultural Implements, and Ploughs
Remarks: They are listed in the 1855 *Maine Registry and Business Directory* (Wood, 1935, 165). In the ephemera collection of Raymond Strout is a receipt for one set of joiner knives dated Sept. 11, 1871 from Hinckley & Egery Iron Co.

Hinckley, Aaron**   Topsham   1817-1840
Tools Made: Blacksmith
Remarks: Aaron assumed his brother Ezechial's business in 1817 and carried it on into the 1840s. In 1828, he occupied a portion of William Whitten's fulling mill at the outlet of the Granny Hole Stream and had a trip hammer, the only one, it is thought, ever used in the vicinity (Wheeler 1878, 610).

Hinckley, Ezechial**   Topsham   1812-1817
Tools Made: Blacksmith
Remarks: His blacksmith business was turned over to his brother, Aaron, in 1817 (Wheeler 1878, 610).

Hinds, Samuel H.   Kingfield   -1869-
Tools Made: Rakes

Hinkley, G. G.**   Augusta   1900
Tools Made: Gunsmith
Remarks: Demeritt (1973, 164).

Hitchcock, Gad**   Yarmouth   1871
Tools Made: Gunsmith
Remarks: Demeritt (1973, 164).

Hobart, D. F.   Madison   -1880-
Tools Made: Rakes, specifically Horse Rakes
Remarks: First name was either Daniel or David.

Hobart, Joel W.   Cornville   -1871-1885-
Tools Made: Rakes (Horse Rakes)

Hobbs, Amos & Sons   Wilton   -1869-1885-
Tools Made: Rakes

Hobby**   Solon
Tools Made: Axes

Remarks: David Smith, coach of the Colby College Woodsmen ax-throwing team has a broad ax marked **HOBBY SOLON**.

Hobby & Parkman　　Solon　　-1849-
Tools Made: Edge Tools

Hobey*　　Solon
Tools Made: Axes

Hobs, J.*　　Norway　　1832-1863
Tools Made: Axes

Hodgen, Cobb　　Gorham　　-1832-1857-
Tools Made: Edge Tools

Hodgkins, E. L.**　　Mount Desert Island(?)
Tools Made: Planes
Remarks: During a visit to the Northeast Harbor Maritime Museum in 2005, a 22 ½" mahogany razee plane and a 9" rosewood smooth plane bearing the mark **E.L. Hodgkins** were noted in the storage area. Both tools had Buck Brother irons and were typical of the boat shop tools made by Maine boat builders using tropical woods obtained in the coasting trade during the 19th century. Hodgkins is a common Mt. Desert Island area name. We have not had time to research the local history of this boat builder. Additional information would be appreciated.

Hodgkins, Eli　　Greene　　-1899-1900-
Tools Made: Farm Tools

Holbrook & Richardson　　Waterville　　1834-1849
Tools Made: Axes

Holbrook, Samuel　　St. Albans　　-1849-
Tools Made: Edge Tools

Holden, Prescott**　　Bangor　　-1855-
Tools Made: Pump and Block Makers
Remarks: He is listed in the 1855 *Maine Registry and Business Directory*.

Holland, C. H.**　　Portland　　-1855-
Tools Made: Brass Founders and Finishers
Remarks: He is listed in the 1855 *Maine Registry and Business Directory*.

Holland, Charles T.   Bangor   -1860-1868
Tools Made: Rules
Remarks: See Norton, Asa H.

Hollis, Brad*   China   1856-
Tools Made: Axes

Holmes, E.**   ?   -1830-
Tools Made: Planes
Remarks: Pollak describes the dado plane in the Bob Jones collection that was found in Maine and has the mark **E.HOLMES** (Pollak 2001, 206). It is unknown whether he bears any relation to the more common I. P. Holmes of Berwick.

Holmes, I. P.   Berwick   ca. 1810 or 1850?
Tools Made: Wood Planes
Remarks: Holmes marked his planes **I.P.HOLMES |
BERWICK ME.** DATM (Nelson 1999) dates these
planes ca. 1850. Pollak reports one example of a 10"
beech molder ca. 1810 (2001, 207). The Bob Jones
collection has a 9 ½" beech round plane.
The Davistown Museum has a 10"
rounding plane (ID# 50402T4) clearly
marked **I·HOLMES** with an owner's mark
of **C REED** and **C.R** (photos) with an early
19th century look (ca. 1810) in the MIII
collection.

Holmes, J.   ?   -1800-
Tools Made: Planes
Remarks: DATM (Nelson 1999) reports one of the J. Holmes marks. Pollak (2001, 207) reports three variations of **J·HOLMES**. The Bob Jones collection has a rounding plane found in Maine with this mark. Bob St. Peters has an 18th century rounding plane marked **J** or **I HOLMES**. Another Holmes mystery.

Holt & Morrill**   Bangor   1887-1896
Holt & Clewley**   Bangor   1900-1908
Holt & Kendall**   Bangor   1909-1913
Tools Made: Gunsmith
Remarks: Demeritt (1973, 164) notes that James W. Holt (d. 1913) was a partner in these three businesses. He worked by himself from 1882 - 1886.

Holt, George   Dixfield   1879-1889
  Tools Made: Gunsmith
  Remarks: Demeritt (1973, 164).

Holt, Hiram & Co.   East Wilton   -1866-1900-
  Tools Made: Farm Tools and Scythes
  Remarks: The lengthy mark: **MANUFACTURED ONLY BY THE HIRAM HOLT CO. | EAST WILTON, FRANKLIN COUNTY, MAINE | WEYMOUTHS PATENT - MARCH 7, 1871**, had a few variations. Sometimes Holt's name alone appeared, and at other

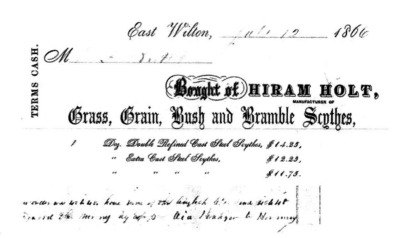

times, "Co." or "& Co." was added. The design for the **LIGHTNING** brand name hay knife he made and sold was patented by George F. Weymouth of Dresden, Maine. He is listed in the 1869 *Maine Business Directory* as Hiram Holt maker, of scythes in E. Wilton and Wilton. The 1882 *Maine Business Directory* listed Hiram Holt & Co. in East Wilton and Wilton. The bill in this figure, courtesy of Raymond Strout, clearly shows a date of 1866. The Davistown Museum has a company file for the Hiram Holt & Co. in vol. 8 of the Hand Tools in History series (Brack 2008b). Also see Holt, Hiram below.

Holt, Hiram   Weld   -1849-
  Tools Made: Farm Tools
  Remarks: This may be the same Hiram Holt as in Hiram Holt & Co. in East Wilton.

Homer, Benjamin H.   Bucksport   -1855-1856-
  Tools Made: Axes and Edge Tools
  Remarks: He is listed in the 1855 and 1856 *Maine Business Directories* as an edge toolmaker and is possibly related to David C. Homer. The Scituate Historical Society has a broad ax marked **B. H. HOMER** that is also marked **STEPHANSON CAST STEEL**, and this second mark does not appear to be an owner's mark. It is unknown if Stephanson was someone working for or with Homer.

Homer, David C.   Bucksport   -1855-1856-
  Tools Made: Axes
  Remarks: It is not known if David worked with Benjamin Homer or was related to him at all. He is listed in the 1855 and 1856 *Maine Business Directories* as an edge toolmaker.

Horton, James C.**   Westbrook   -1869-
Tools Made: Shipsmith
Remarks: He is listed in the 1869 *Maine Business Directory*.

Houghton, Richard W.**   Norway, Mechanic Falls   1838-1863   (d.1863)
Tools Made: Gunsmith
Remarks: Demeritt (1973, 53-5, 164).

Howard Axe Co.*   Island Falls   1899-1914
Tools Made: Axes
Remarks: Lamond (2007b) speculates this and Island Falls Edge Tool may have been the same company.

Howard Mfg. Co.   Belfast   1865-1876
Tools Made: Carpenter Tools
Remarks: According to the DATM (Nelson 1999), the information on this company is confusing and contradictory. "Supposedly, Franklin Augustus Howard and Hollis M.A. Poor made a miter machine patented by Howard" as well as proof presses. "F.A. Howard continued alone as a screwdriver maker after 1876." The DATM suggests that Howard may have continued to be a part of this company after he began to make screwdrivers under his own name. See Howard & Son, F. A. and Howard, Franklin Augustus.

Howard, Caleb**   Waldoboro   -1778-
Tools Made: Blacksmith
Remarks: The early inhabitants of Union, having no blacksmith, would employ Howard. Howard would bring ox-shoes, nails, anvil, and hammer. In December of 1778, he was set up in Philip Robbins' newly constructed barn. The sparks flying into the hayloft ignited the barn. The Robbins families' grain and hay reserves were destroyed (Sibley 1851, 44-5).

Howard, F. A. & Son   Belfast   1895-1901-
Tools Made: Screwdrivers
Remarks: The mark stamped on the screwdrivers was: **Made by | F.A. Howard & Son | Belfast, ME | U.S.A.** Sometimes the U.S.A. part was on the same line as the town and state, and either of the patent dates Mar. 1 1892 or July 23, 1895 followed. F. A. Howard had been making spiral screwdrivers under his own name until 1895, when his son, William Russell Howard, became his partner. Two different patents of the same date (1 March 1892) were issued to F. A. Howard and J. W. Jones and both were used by the company, as well as a 23 July 1895 patent, whose recipient is unknown. The Museum has two Howard & Son screwdrivers (ID# 111001T25 and 121600T2) in the IR collection. See

Howard, Franklin Augustus, Howard Mfg. Co., and Clifford Fales' (1992) *Gristmill* article on F. A. Howard spiral screwdrivers.

Howard, Franklin Augustus   Belfast   1876-1895
Tools Made: Dies, Screwdrivers, and Adjustable Mitering Machine
Remarks: Howard was primarily known for his spiral screwdrivers, for which he held one patent (1 March 1892) and he manufactured screwdrivers patented by other inventors. Those included two by Isaac Allard (patented 4 Aug. 1868 and 24 Nov. 1874) and one by J. W. Jones (patented 1 March 1892). Another patent date (23 July 1895), whose inventor is unknown, was also manufactured. He marked his tools: **F.A. Howard | -Maker- | Belfast, Me.** (along with the patent date and holder's name), until 1895 when the name changed to **F.A. Howard & Son**. See Howard & Son, F. A. and Howard Mfg. Co.

Howes, J. R.**   Orland   -1856-
Tools Made: Shipsmith
Remarks: He is listed in the 1856 *Maine Business Directory*.

Howland, James   -1879-
Tools Made: Plows

Hubbard & Blake   Waterville   1862-1865
Tools Made: Axes and Scythes
Remarks: Later John U. Hubbard and William P. Blake became Hubbard, Blake & Co. and, then, Hubbard & Blake Mfg. Co. Also see H & B Mfg. Co.

Hubbard & Blake Mfg. Co.   Oakland, Waterville   1879-1889
Tools Made: Axes and Scythes
Remarks: This company is listed in the 1879, 1881, and 1882 *Maine Business Directories* as located in West Waterville, which became Oakland. The company is listed later in Waterville. It was sold to the American Axe & Tool Co. See Hubbard & Blake.

Hubbard, Blake & Co.   Waterville   1865-1877
Tools Made: Axes and Scythes
Remarks: This company, comprised of John U. Hubbard, William P. Blake, Luther D. Emerson, and Charles E. Folsom succeeded Hubbard & Blake only to return to be Hubbard & Blake Mfg. Co. when Emerson and Folsom split off to form Emerson & Stevens. Although they left in 1870, the name change did not occur until 1877. This company is listed in the 1867 and 1869 *Maine Business Directories* as located in West Waterville and Waterville.

Hubbard & Matthews   Waterville   1854-1858

Tools Made: Axes and Scythes

Remarks: This partnership was listed for the above years as scythe makers in a Waterville directory. The DATM (Nelson 1999) states "An obscure mark on an axe has the same name Mathew(s?) in a different letter style than a Hubbard (& Co.?) and Waterville." Also John U. Hubbard of Waterville was later known as an ax and scythe maker in several companies. It is unknown if this is an earlier venture of his or a relative.

Hunnewell & Hill**   Moscow   -1879-

Tools Made: Edge Tools

Remarks: This company is listed in the 1879 *Maine Business Directory*. See Em. Hill and Able J. Hunneywell, possibly this was a partnership of those two men.

Hunneywell, Able J.*   Moscow   1878-1887

Tools Made: Axes

Remarks: He is listed in the 1881 and 1882 *Maine Business Directories*. Hoyt (1881) also lists Abel Hunnewill of Moscow as a smith.

Hunnywell & Ellis*   Kingfield   1896-1904

Tools Made: Axes

Hunt, W. H.*   Liberty   -1874-

Tools Made: Axes

Hunter, C.**   Bingham

Tools Made: Axes

Remarks: The Davistown Museum has a broad ax (ID# 100605T3) marked **C HUNTER BINGHAM** in its MIII collection. The first initial is hard to read and may not be C. This tool is part of the "Art of the Edge Tools" exhibition.

Hunter, D. G.**   Camden   -1881-1882-

Tools Made: Edge Tools

Remarks: He is listed in the 1881 and 1882 *Maine Business Directories*. Lincolnville is quite close to Camden, so this might be the David Hunter of 1856.

Hunter, David   Lincolnville   -1856-

Tools Made: Edge Tools

Remarks: He is listed in the 1856 *Maine Business Directory*.

Huntington, James**   Portland   1866

Tools Made: Gunsmith

Remarks: Demeritt (1973, 164).

Huntley, Abiel**   Houlton   1876-1880
Tools Made: Gunsmith and Jeweler
Remarks: Demeritt (1973, 164).

Hurd, William   Liberty   -1862-1896-
Tools Made: Axes and Edge Tools
Remarks: He is listed in the 1867, 1869, 1881, and 1882 *Maine Business Directories*.

Hurman, Nathan   Buxton   -1828-1860-
Tools Made: Edge Tools

Huson, E. L.**   Machiasport   -1855-
Tools Made: Pump and Block Makers
Remarks: He is listed in the 1855 *Maine Registry and Business Directory*.

Hussey Mfg. Co.   North Berwick   1835-1971
Tools Made: Farm Tools
Remarks: "Established" in 1835, it may have operated under a different name before 1900 (Nelson 1999).

Hussey, Timothy B.   North Berwick   -1869-1885-
Tools Made: Farm Tools and Plows, including Cultivators
Remarks: He may have a connection to Hussey Mfg. Co. also of North Berwick.

Hutchins, Isaac   Wellington   -1879-
Tools Made: Farm Tools

Hutchins, Ivory**   York   -1870-
Tools Made: Gunsmith
Remarks: Demeritt (1973, 164).

Hutchinson, W. P.**   Yarmouth   -1869-
Tools Made: Shipsmith
Remarks: He is listed in the 1869 *Maine Business Directory*.

Hyde, Thomas W.   Bath   -1869-
Tools Made: Farm Tools

Ingalls, Brown**   Bangor, Blue Hill, Bucksport, Portland   1846-1889
Tools Made: Gunsmith

Remarks: Demeritt (1973, 164) notes he moved from Bangor to Blue Hill in 1856, then to Bucksport in the 1860s, to Portland in 1880, and finally to Stewardson, Illinois in 1889.

Ingalls, Frank**   Eastbrook   -1882-
Tools Made: Edge Tools
Remarks: He is listed in the 1882 *Maine Business Directory*.

Ingalls, J. M.**   Bath   -1855-
Tools Made: Pump and Block Makers
Remarks: He is listed in the 1855 *Maine Registry and Business Directory*.

Ingalls, William**   Bath   -1854-
Tools Made: Blocks
Remarks: Baker (1973, 436) in *A Maritime History of Bath, Maine* lists William Ingalls as one of three block-makers in Bath.

Iones, E**   -1790-1810-
Tools Made: Planes
Remarks: Pollak (2001, 222) states that the last name is probably "Jones." Examples, marked **E: IONES** include a 14 1/2" crown molder struck twice on the toe, with birch offset tote, a 13" double iron sash with a birch tote, a 9 3/8" complex molder, a plow with a riveted skate, a 9 1/2" hollow, and a 9 1/2" double bead found in Maine. Items are all beech with heavy flat chamfers, with the exception of the crown, which has round chamfers.

Island Falls Edge Tool*   Island Falls   1902-1914
Tools Made: Axes
Remarks: Lamond (2007b) speculates this and Howard Axe Co may have been the same company.

J. D. (see J. Dearborn)

Jackson, Edward T.**   Bangor   1834
Tools Made: Gunsmith
Remarks: Demeritt (1973, 164).

Jackson, S. R.   Foxcroft   -1871-
Tools Made: Farm Tools

Jackson, T. C.   Bath   -1855-1869-
Tools Made: Adzes and Chisels

Remarks: Jackson marked his tools **T.C. JACKSON |
BATH.ME.** with the initial "C" looking very much like an
"O". Yeaton (2000) notes that T. C. Jackson was also an ax-
maker during the time period of 1832-186. He is listed in the
1855 and 1856 *Maine Business Directories* as an edge
toolmaker. The Davistown Museum has a Jackson
shipbuilder's adz (ID# 42403T1) in the MIV collection (photo).

Jameson, Samuel** Topsham 1836-1873
Tools Made: Blacksmith
Remarks: He was in business with James Maxwell as Maxwell & Jameson.

Jaquith, Chas.** Clinton -1882-
Tools Made: Axes and Edge Tools
Remarks: He is listed in the 1882 *Maine Business Directory*. The Davistown Museum
has an advertisement for "Chas. Jaquith general Blacksmith, Manufacturer of Axes and
all kinds of edge tools." It is from the Clinton Advertiser and is accompanied by an ad for
John P. Billings of Clinton.

Jennings, Solomon** Richmond, Fort Halifax (now Winslow) 1752-1756
Tools Made: Gunsmith
Remarks: Demeritt (1973, 164) notes he moved from Richmond to Fort Halifax in
1754.

Jewett, J. C. Waterville ca. 1820
Tools Made: Wood Planes
Remarks: Pollak (2001, 221) states that a number of planes have been reported,
including a jack, bead, hollow, sash, and side rabbet. A John C. Jewett, occupation
unknown, was reported living in Waterville in the census of 1830. His planes were
marked: **J.C.JEWETT | WATERVILLE** or just **J.C.JEWETT**. The Bob Jones collection
contains at least three examples of his planes. The Davistown Museum has a Jewett sash
plane in the MIII collection. In 2007, a large collection of Jewett's planes was dispersed
by the Liberty Tool Co.

Jewett, J. M. Bangor -1869-1871-
Tools Made: Files
Remarks: Jewett was the proprietor of Bangor File Manufactory before he advertised
files under his own name.

Jewett, Nathan   Aurora   -1869-1871-
Tools Made: Axes
Remarks: He is listed in the 1869 *Maine Business Directory*.

Johnson, H. & Co.   Bloomfield   1846-
Tools Made: Handles, specifically Shovel Handles

Johnston, Francis**   Waldoboro   -1856-
Tools Made: Shipsmith
Remarks: He is listed in the 1856 *Maine Business Directory*.

Jones & Hunt**   Brunswick   -1825-
Tools Made: Blacksmith

Jones, Asa**   Portland   1821-
Tools Made: Edge Tools, Blacksmith, and
Whitesmith
Remarks: The illustration is of an advertisement
that was in the *Independent Statesman*. It indicates the
shop was previously owned by Daniel Gilbert and
employs Stenchfield and Tufts.

Jones, Alphonzo C.**   South Paris   1879-1900
Tools Made: Gunsmith and Machinist
Remarks: Demeritt (1973, 164).

Jones, Benjamin L.**   Union   -1855-1856-
Tools Made: Shovels and Spades
Remarks: He is listed in the 1855 *Maine Registry*
and 1856 *Business Directory*.

THE INDEPENDENT STATESMAN
PORTLAND, M.E. AUG. 11, 1821

all who may be pleased to favor the above establishment.

Portland, July 14, 1821.   tf.

ASA JONES,
Black and White Smith,
MIDDLE-STREET......PORTLAND,

RESPECTFULLY informs the inhabitants of this town and vicinity, that he has taken the shop formerly occupied by Mr. DANIEL GILBERT, and latterly by Messrs. *Stenchfield & Tufts*, where he intends carrying on the

BLACK & WHITE SMITH
BUSINESS,
in all its various branches.

Warranted EDGE TOOLS; IRON MACHINERY, and Sheet Iron Work, may be had as above.

☞ ALSO, *Horse Shoeing and Carriage Work*, executed in superior style.

*** Messrs. STENCHFIELD & TUFTS, are now employed in the above shop.

August 11.

Jones, D. J.**   Durham   -1867-
Tools Made: Edge Tools
Remarks: He is listed in the 1867 *Maine Business Directory*.

Jones, David & Son   Woodstock   -1862-1866-
Tools Made: Edge Tools

Jones, E (See E. Iones)

Jones, G. H.   Union
Tools Made: Farm Tools

Jones, J. W.   Belfast   -1892-1902-

Tools Made: Screwdrivers

Remarks: Holding several patents for screwdrivers (1 March 1892 and 23 July 1895), which F. A. Howard is known to have manufactured, Jones may have also been employed with Howard's company. Either way, Howard did not have sole rights to the patents, as they have been noted on other screwdrivers under Jones' name only. Whether Jones was producing them on his own or if another company was using his patent is not known. According to C. D. Fales (1992) he held patent 470005, 3/1/1892 with Frank Howard; patent 543096, 7/23/1895.

Jones, James**   Brunswick   -1810-

Tools Made: Blacksmith

Jones, L. M.**

Tools Made: Planes

Remarks: Pollak (2001, 224) states that three planes found in Portland, Maine, were marked inside an oval **L.M. JONES BEST** with a star in the center. All were marked with an owner imprint of **E. E. Bowden**, a shipwright ca. 1850. The Bob Jones collection has a plane with this mark that was found in Maine, a coffin smooth, 8 ¾", with a lignum vitae body and an iron blade with chip breaker marked **CHARLES BUCK WARRANTED CAST STEEL.**

Jones, R.**

Tools Made: Planes

Remarks: Pollak (2001, 225) lists two planes with an **RJONES** mark. Bob Jones states that the 7 5/8" smooth plane marked **R · JONES** in his collection could be the Robert L. Jones that Pollak lists from Nova Scotia or possibly Rockland Llewellen Jones, who was a finish carpenter and may have made or marked his own tools.

Jones, S. W.**   South Union

Remarks: He owned an iron foundry. An advertisement for it is in the *Union Weekly Times* (1895) history of Union, Maine.

Jones, T. C.**   Solon   1880-1881

Tools Made: Smith

Remarks: He is listed in Hoyt (1881).

Jones, William Trebey**   Augusta   1882-1890-   (d. 1926)

Tools Made: Gunsmith

Remarks: Demeritt (1973, 164).

Jordan, James* Cherryfield 1855-1858
Tools Made: Axes

Jordan, Samuel** Bath -1881-
Tools Made: Shipsmith
Remarks: He is listed in the 1881 *Maine Business Directory*.

Jordan, W.* Waterville 1854-1857
Tools Made: Axes

Judkins, Robert F. North New Portland -1849-
Tools Made: Edge Tools

Junkins, Charles North Berwick -1862-
Tools Made: Edge Tools

Kavanagh, Frank** Brewer -1882-
Tools Made: Shipsmith
Remarks: He is listed in the 1882 *Maine Business Directory* on Main St.

Keene, C.** -1830-
Tools Made: Planes
Remarks: Pollak (2001, 227) lists the following examples, marked **C. KEENE**, a 7 1/2"
beech wedge-arm plower plane, two 9 7/16" fruitwood skew rabbets, one with a **VII** on
the wedge and iron, both found in Maine, and a S. Cumings complex molder with
shallow round chamfers.

Keene, Galon Appleton -1879-
Tools Made: Rakes, including Horse Rakes

Keith, J. F. Buckfield -1885-
Tools Made: Farm Tools, including Cultivators

Keith, John** Mars Hill -1881-
Tools Made: Axes and Chisels
Remarks: He is listed in the 1881 *Maine Business Directory*.

Kelley, Benjamin & Co. Belfast -1855-1929
Tools Made: Axes and Edge Tools
Remarks: Kelley used the mark **B.KELLEY&CO | BELFAST, MAINE**. Benjamin Kelley
is listed in the 1855 and 1867 *Maine Business Directories*. The 1879 *Maine Business
Directory* lists Benjamin Kelley & Co. The 1881 *Maine Business Directory* lists B.

Kelley & Co. of East Side, Belfast. It also lists B. Kelley separately. It is unknown if this company kept changing names. Lamond (2007b) gives the end date of 1929. See Benjamin Kelley & Sons below. The Davistown Museum has a Kelley slick, ID# 040904T1 (photo) and a mast ax in the MIV collection. Dana Phillipi has a hewing ax marked "Kelley Belfast Maine cast steel".

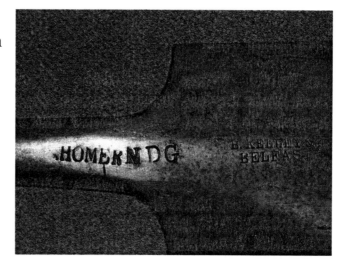

Kelley, Benjamin & Sons   Belfast -1869-1871-
   Tools Made: Axes
   Remarks: "Sons" is used in a 1869 directory, while only "Son" is used in 1870.

Kelley, B.**   Bangor   -1881-
   Tools Made: Shipsmith
   Remarks: He is listed in the 1881 *Maine Business Directory* at 8 Washington St.

Kelley, Daniel T.   Portland   -1883-1885-
   Tools Made: Farm Tools and Plows, including Mowing Machines
   Remarks: In the ephemera collection of Raymond Strout is a letter on Daniel T. Kelley stationary noting the production of "The New Model Centre Draft Mower" dated Oct. 8, 1883.

Kelley, E. J.**   Bangor   -1882-
   Tools Made: Shipsmith
   Remarks: He is listed in the 1882 *Maine Business Directory* at 12 Washington St.

Kelley, E. W.   Winthrop   -1849-
   Tools Made: Farm Tools

Kelley, James   Cherryfield   -1855-1882-
   Tools Made: Axes and Edge Tools
   Remarks: His name has also been seen under the "Kelly" spelling, where Yeaton (2000) lists the working dates as 1855-1858. He is listed as Kelly in the 1855 *Maine Business Directory* as an edge toolmaker. He is listed in the 1867, 1869, 1879, 1881, and 1882 *Maine Business Directories* as Kelley.

Kelley, Manley   Mt. Vernon   -1880-1885-
Tools Made: Farm Tools and Rakes, including Drags

Kendall & Whitney*   Portland
Tools Made: Axes

Keyes, Calvin   Wilton   1838-1869-   (b. 1814 d.1864)
Tools Made: Scythes, Drawknives, and Edge Tools
Remarks: He is listed in the 1855 and 1856 *Maine Business Directories*. According to W. A. "Chet" Sweatt, in 1839 he became a partner in Andrew Butterfield's blacksmith shop in East Wilton. Hiram Holt took title to the shop from the Keyes estate in 1864. DATM (Nelson 1999) gives his working dates as 1855 - 1869, which seems to conflict with the death date. The Wilton Museum has a scythe blade with a paper label: **SUPERIOR SCYTHES, | Made expressly for Retail Trade, by | CALVIN KEYES, EAST WILTON, ME.** Sweatt had in his collection a 10" drawknife marked **CAST STEEL | C. KEYES, | E. WILTON, | WARRANTED.**

Kidder, W. L.**   -1810-1820-
Tools Made: Planes
Remarks: Pollak (2001, 234) lists the following examples, marked **W.L. KIDDER**, a 10 1/2" cherry shiplap rabbet with a closed integral tote and heavy flat chamfers and a 22" beech jointer with round chamfers found in Maine.

Kierstead & Barker**   Danforth   -1881-
Tools Made: Axes
Remarks: This company is listed in the 1881 *Maine Business Directory*. Possibly this is Charles Kierstead and Daniel Barker of nearby Weston.

Kierstead, Charles**   Danforth   -1879-
Tools Made: Edge Tools
Remarks: He is listed in the 1879 *Maine Business Directory*.

Kimball & Sanford   Bangor   -1872-
Tools Made: Saws
Remarks: Saw makers and dealers in a range of tools, Reuel W. Kimball and John E. M. Sanford were connected to "Dirigo Saw Works" and "Gibson, Kimball & Sanford," as implied by their 1872 invoice that shows a circular saw bearing those two names. A company that included Gibson was superceded by this partnership. See Reuel W. Kimball.

Kimball, D. S.   Minot   -1849-
Tools Made: Edge Tools

136

Kimball, David   Amherst   -1849-
   Tools Made: Edge Tools

Kimball, Joel*   Buxton   -1874-
   Tools Made: Axes

Kimball, Reuel W.**   Bangor   -1876-
   Tools Made: Saws
   Remarks: R. W. Kimball issued a price list of saws, including "Henry Disston & Sons Celebrated Saws," anti-friction metals, shingles, lathes, and machinery dated February 14, 1876. His flyer indicated he was a successor to Kimball & Sanford, possibly only as a vendor.

King & Messer*   Oakland   1904-1907
   Tools Made: Axes

King, Benjamin Jr. & Peter   Whitefield   -1856-
   Tools Made: Axes
   Remarks: Both Peter King and Benjamin King Jr. are listed in the same 1856 directory, but the connection, if there was one, is not known. Whether they worked together or separately is also unknown. Peter was a particularly important edge toolmaker who was born in 1804. His grandfather, Benjamin King, lived in Kings Mills, Whitefield, Maine circa 1790 and died in 1801. It is unknown who Benjamin King Jr. was.

King, J. H.**   Portland   -1855-
   Tools Made: Coffee and Spice Mills
   Remarks: He is listed in the 1855 *Maine Registry and Business Directory*.

King, J. K.
   Tools Made: Planes
   Remarks: Pollak (2001) states "Example: a 11" birch closed toted rabbet, a 9 5/16" beech skew rabbet, and a 9 ½" beech side rabbet, all with round chamfers. The side rabbet also has the incuse imprint branded on the side and was found in ME with two planes by S. King (w.s.) ca. 1820-30."

King, John   Bangor, Oakland

Tools Made: Axes and Scythes

Remarks: Axes bearing the same maker name, but different towns (Oakland, ME and Bangor, ME), have led to speculation as to whether there were two men or just one. It is unknown if the John King in Oakland later become the John King Axe Co. (1907-17) also located in Oakland. Still another possibility lies in Canada, where an ax-maker with the name John King was reported both in Ontario and Quebec around 1887 - 1892. The Museum has an ax (ID# 12801T12, photo of ax head and mark) signed **JOHN KING | OAKLAND, ME.** in the IR collection. Tom Lamond (personal communications) notes that John King of Oakland has Pat. No. 615,518 for a scythe, Dec. 6, 1898. It is shared with Sanford J. Baker, also of Oakland.

King, John Axe Co.   Oakland   1877-1927

Tools Made: Axes

Remarks: The mark: **JOHN KING AXE CO. | OAKLAND, MAINE** is printed on a label and not imprinted on the ax itself. Thought to have begun shortly after 1877, the company is followed by a King Axe & Tool Co. working out of Oakland, which appears in the registries 1907-17 and may have been a successor. Yeaton (2000) lists King Axe & Tool Co. working from 1908-1927. The photograph (bottom right in group above) is of an ax with a knife concealed by screwing it into the handle. It is one of only four known examples and is believed to have been made by King Axe Co. of Oakland but is not marked. Photograph courtesy of Rick Floyd.

King II, Peter*   Whitefield   -1856   (b.1804 d.1858)

Tools Made: Axes

Remarks: An ax marked **KING P.K.** has been reported to the museum and is in the collection of Roger Majorowicz of Iron Horse Sculpture, who supplied this photograph. Peter King II was a particularly important edge toolmaker who was born in 1804 and had his blacksmith shop on the Sheepscot River below his home in Whitefield. The area where his foundry was located came to be known as Kings Mills, see *King's Mills, Whitefield, Maine 1772-1982* by Henry Waters (1982), which contains an extensive 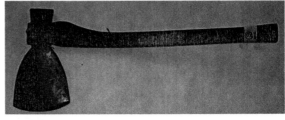 summary of the King family beginning with Benjamin King, Sr., who first bought land from Abraham Choate in 1790 and later his house and mill in 1801. He was killed in an accident at the mill the same year. The property then descended to Peter King I, who may have been the father of Peter II. It is this second Peter who is the famous edge toolmaker. After his death, his son, Sauren King, continued as a blacksmith at this location. A number of Peter King II edge tools are in the collection of the sculptor, Roger

Majorowicz, who now lives on the King property. See Benjamin Jr. & Peter King and Linwood Lowden's (1984) *Ballstown West - 1768 - 1809*.

King, S.** ? ca. 1720-1820
Tools Made: Planes
Remarks: Pollak (2001) states "Examples: the molders are 9 ¼" - 9 7/8" birch and beech, with flat and round chamfers, indicating a transitional planemaker; a 10" beech Yankee plow with wedge arm lock, wood depth stop, screw locked; and a 29" beech bench plane with single irons, center closed tote and 19c style chamfers. Two have been found with J. K. King planes from ME, suggesting a possible connection. The initial imprint has the same construction details, wedge profile and imprint design as the full name. It appears on a 9 ½" beech skew rabbet with heavy round chamfers. ca. 1780-20, probably from New England. S. King is not to be confused with S. King of Hull, England." The marks described are: **S.KING** and **S.H.K.** The Davistown Museum has an S. King 9 ½" birch Yankee plow plane (ID# 111104T1) in the MII collection.

King, Samuel** Paris (d. 1856)
Tools Made: Apple Parer
Remarks: The following information was provided by Samuel King's great great grandson, Lincoln King. Lincoln has in his possession an apple parer and a fine desk made by Samuel King, who was a carpenter and builder. Captain Samuel King is listed in the *History of Paris* by William Lapham (1884, 651). It states that he was the "son of Sergeant George, [and] came to Paris with his uncle, Jairus Shaw, settled first on High street, and afterward exchanged farms with Asa Barrows, now the homestead of William O. King." His son Horatio King became the Postmaster General of the US under President Buchanan.

King, Sorren W.** Whitefield -1855-
Tools Made: Edge Tools
Remarks: He is listed in the 1855 *Maine Business Directory*. It is unknown if Sorren was a member of the King clan of ax-makers in Whitefield.

Kittredge, John H.** Augusta 1871-1890- (d. 1928)
Tools Made: Gunsmith and Locksmith
Remarks: Demeritt (1973, 164).

Kneeland, G. S. Lincoln -1867-1871-
Tools Made: Axes and Edge Tools
Remarks: He is listed in the 1867 and 1869 *Maine Business Directories*.

Knight, F. C.   Bridgton   -1899-1900-
   Tools Made: Farm Tools

Knight, Samuel C.   Cornish   -1832-1871-
   Tools Made: Axes and Edge Tools
   Remarks: The 1832 date found in one source seems a bit early for a single maker who was cited in several directories around 1870. However an S. C. & J. Knight is listed in a Cornish directory (unfortunately of an unspecified year) (Nelson 1999). He is listed as making axes in the 1867 and 1869 *Maine Business Directories*.

Knowles, Graues**   Falmouth   1722
   Tools Made: Gunsmith
   Remarks: Demeritt (1973, 164).

Knowlton & Merryman**   Brunswick   -1845-
   Tools Made: Blacksmith

Lamb, Andrew**   Robbinston   -1856-
   Tools Made: Shipsmith
   Remarks: He is listed in the 1856 *Maine Business Directory*.

Lamb, Luther R.**   Waterville   1846
   Tools Made: Gunsmith
   Remarks: Demeritt (1973, 164).

Lambard, C. A. & Company**   Bath   -1854-
   Tools Made: Ships' castings
   Remarks: An iron foundry. "Its work was confined mainly to the manufacturing of ships' castings the use of which in recent years had grown enormously as iron replaced wood for many items on shipboard" (Baker 1973, 434).

Lambert Nail Holder Mfg.**   Phillips   -1910-
   Tools Made: Nail Holders
   Remarks: On March 9, 1910, this company patented a detachable nail holding device that clips on to a hammer handle. It would be used to help start the nail. Apparently, this idea did not catch on. (Photo courtesy of Rick Floyd.)

Lamont, Alfred (See Alfred Lemont)

Lancaster, B. F.**   Augusta   1885
   Tools Made: Wrenches

140

Remarks: Herb Page has the adjustable buggy wrench pictured here. It is stamped **American Wrench Co. Augusta Me** and with the **September 8, 1885** patent date.

Lancaster, H. S.** Ellsworth 1873-1900
Tools Made: Gunsmith and Blacksmith
Remarks: Demeritt (1973, 164).

Lancaster, J. W.** Ellsworth 1876-1894
Tools Made: Gunsmith and Blacksmith
Remarks: Demeritt (1973, 164).

Landers, J. C. Gardiner -1899-1900-
Tools Made: Farm Tools

Lane, Ellsworth S.** Upton -1942 (1877b. - 1945d.)
Tools Made: Rules, Log Calipers, and Log Scales
Remarks: See the publication by Butterworth and Blumenberg (1993) "E. S. Lane: A Maine rule maker and scaler" in *The Chronicle.*

Larned & Hale Waterville 1836-1839
Tools Made: Scythes
Remarks: The name was changed to S. & E. Hale after Larned sold out his interest.

Larrabee, Benjamin** Porter -1855-
Tools Made: Edge Tools
Remarks: He is listed in the 1855 *Maine Business Directory.*

Larrabee, J. C.** Brunswick
Tools Made: Planes
Remarks: Bob Wheeler has a panel plane signed by Larrabee, circa 1780 - 1820. The Museum has a Larrabee skew plane (ID# 50402T2) in the MII collection (photo).

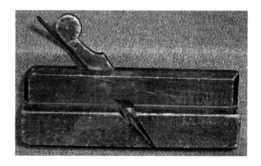

Laughlin, Robert** Robbinston -1856-
Tools Made: Shipsmith
Remarks: He is listed in the 1856 *Maine Business Directory.*

Laughlin, T. Co. Portland -1869-1882-
Tools Made: Chisels, specifically Caulking Chisels/Irons, Shipsmith
Remarks: Their chisels were marked **T. LAUGHLIN Co / PORTLAND, ME.** The *Maine*

Business Directory in 1869 lists Thomas Laughlin & Son, 185 Commercial St, Portland, in 1881 at Commercial, cor. Centre, and 18 and 20 Centre, and, in 1882, at Commercial, cor. Centre. The Davistown Museum has a T. Laughlin caulking iron (ID# TCX1005) in the IR collection.

Lawrence, William**   Farmington
Tools Made: Gunsmith
Remarks: Demeritt (1973, 83-4, 164). See Wheeler & Lawrence.

Leach, S. F. & Co.   Bangor   -1869-
Tools Made: Saw Tools
Remarks: S. F. Leach held a 19 Jan. 1869 patent for a saw set, which he made under his own name in Maine but then adopted the "& Co." once he was working in Boston, MA., and then later became part of Leach & Towle. Leach's saw set was also being made by O. W. Bullock & Co, and, in 1901, C. E. Jennings & Co. was selling it, perhaps even making it.

Leavitt, C. F.**   Lewiston   -1879-1882-
Tools Made: Stone Tools
Remarks: He is listed on Bates St. in the 1879 and 1881 *Maine Business Directories* and at 3 Franklin in the 1882 *Maine Business Directory*.

Leighton, E. C.   Winthrop
Tools Made: Cooper Tools and Metal Planes, including Coopers' Planes
Remarks: He used the mark: **E.C. LEIGHTON | WINTHROP, ME.** Pollak (2001, 249) reports a brass and wood coopers' croze in the Bob Jones collection.

Leighton, Palmer**   Cherryfield   -1856-
Tools Made: Shipsmith
Remarks: He is listed in the 1856 *Maine Business Directory*.

Leland, Henry**   Augusta   1840-1846
Tools Made: Gunsmith
Remarks: Demeritt (1973, 164).

Leland, Henry   Sedgwick   -1849-
Tools Made: Edge Tools

Leland, Larkin M.**   Augusta   1838-1894   (d. 1894)
Tools Made: Gunsmith
Remarks: Demeritt (1973, 79-82, 164).

Lemont, Alfred**   Bath   -1830-1835-   d. 1896
Tools Made: Shipsmith
Remarks: Alfred Lamont built the schooner *Eliza Ann* in Bath in 1835 and then went into business with James F Trott as shipsmiths, providing the Bath region with "hoops, anvils, vises, crowbars, plow molds, mill saws, fisherman's anchors, and all sorts of edge tools" (Baker 1973, 429). See Trott & Lemont. He spent a 15 year period in the shipyard of Richard Morse & Sons in Phippsburg. In 1851, he began shipbuilding on his own in partnership with William M. Reed as Reed & Lemont and later with master builder Alexander Robinson.

Leonard, Austin   Windham   -1867-1871-
Tools Made: Edge Tools
Remarks: He is listed in the 1867 and 1869 *Maine Business Directories*.

Leonard, Hiram L.**   Bangor   1854-1880   (d. 1907)
Tools Made: Gunsmith
Remarks: Demeritt (1973, 164) notes he moved to Central Valley, NY in 1880.

Lepens, Frd**   -1800-1810-
Tools Made: Planes
Remarks: Pollak (2001, 250) states that this example, marked **Frd Lepens** comes from a group of 10 1/8" - 9 1/4" beech hollows and rounds with relieved wedges and flat chamfers found in Maine. All are also imprinted with an owner's mark **A. NASH**, which could help in locating this maker.

Levanseller, Ludlow**   Waldoboro   1956-1869-
Tools Made: Shipsmith
Remarks: He is listed in the 1856 *Maine Business Directory*. The 1869 Business Directory lists an L. L. Levensaler as a shipsmith in Waldoborough. It is unknown if this is the same person.

Lewis, Jabez   China   -1871-1879-
Tools Made: Plows

Libby**   Gardiner   +/- 1860
Remarks: The Davistown Museum has recently located an open end wrench, 6" long, (1/2" + 5/8"), which is drop-forged and also inscribed with an E in a circle and the numbers 7338A. It is marked **LIBBY | GARDINER, ME**.

Libby & Bolton   Portland   -1857-1886-
Tools Made: Adzes, Axes, Chisels, Drawknives, and Edge Tools

Remarks: John W. Libby, born 1831, and Elbridge G. Bolton, born 1823, marked their tools several ways: **I.W.LIBBY** | **E.G.BOLTON** | **PORTLAND** and **LIBBY & BOLTON** | **PORTLAND** (sometimes leaving out the city line). There may be a connection with

Higgins & Libby. This company is listed in the 1867, 1869, 1879, and 1881 *Maine Business Directories* with a location of 238 Fore St. The 1882 *Maine Business Directory* gives the address as 460 Fore St. The Davistown Museum has a drawknife (ID# 61204T3), socket chisel (ID# 42604T2, photo) and gouge (ID# TBC1002) made by this company in the MIV collection. Vaughn Kelly has a coopers' jigger made by them.

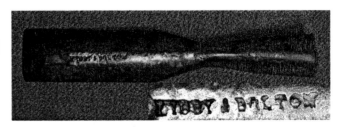

Libby, David   Buxton   -1827-1850-
   Tools Made: Edge Tools

Libby, E.**   -1790-
   Tools Made: Planes
   Remarks: Pollak (2001, 251) lists a 10 1/2" beech skew rabbet with flat chamfers marked **E.LIBBY**. This example is probably from Maine, as Libby is a common name there.

Libby, Frank*   Bangor   -1902-1917
   Tools Made: Axes
   Remarks: He worked in Bangor until 1902, then worked in Portland from 1903 to 1917 (Yeaton 2000).

Libby, H. L.**   Norway   1884
   Tools Made: Gunsmith
   Remarks: Demeritt (1973, 164).

Libby, John F.**   Prospect   -1840-1880-
   Tools Made: Blacksmith
   Remarks: He apprenticed with Charles Turner while working at the quarries. With time, he purchased Mr. Turner's business. For 45 years, he carried on the trade of blacksmith and farrier. He also worked as blacksmith for the quarries (*Biographical Review* 1897, 73-4; Ellis 1980, 251-2).

Libby, John F., Jr.**   Prospect   -1855-
   Tools Made: Blacksmith
   Remarks: He learned the trade from his father and continued working as a journeyman

until injured in an accident when he had to give up the trade (*Biographical Review* 1897, 73-4; Ellis 1980, 251-2).

Liberty Machine Co.**   Liberty

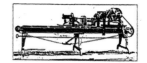

THE LIBERTY TONGUE AND GROOVE STAVE MACHINE

Tools Made: Stave and Heading Machinery

Remarks: The illustration of the Liberty tongue and groove stave machine is from a brochure published in Donahue (1996). Frank Bennett held the patent for this machine (Donahue 1996). See F. P. Bennett & Co.

SIMPLE---DURABLE---ECONOMICAL

CAPACITY: 15,000 STAVES DAILY

MANUFACTURED BY

THE LIBERTY MACHINE CO.
Builders of Stave and Heading Machinery
LIBERTY . MAINE

Licett, James   Springfield   -1869-1871-

Tools Made: Axes

Remarks: He is listed in the 1869 *Maine Business Directory*.

Lie-Nielsen Toolworks Inc.**   Warren   1981-present

Tools Made: Planes and Saws

Remarks: The Lie-Nielsen tool company has been in business for over two decades and is located on Route 1 in Warren, ME, just south of Thomaston. The company produces smoothing planes, block planes, bench planes, shoulder planes, skew block planes, beading tools, chisel planes, scraping planes, and handsaws. Lie-Nielsen is the only production plane company now operating in Maine. Store hours and

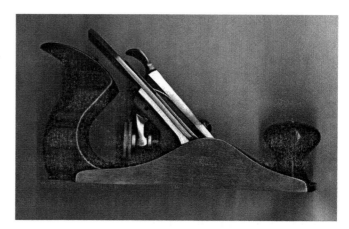

company history are on their website. The Davistown Museum has a Lie-Nielsen (ID# 103102T1, photo) no. 1 size special edition bench plane in the IR collection.

Lilly, Robert**   Bath   -1849-1882-

Tools Made: Shipsmith

Remarks: "In 1849 Robert 'Uncle Bob' Lilly of Dresden, then 19, ...was on his way to learn the blacksmith's trade in Bath. [David] Crooker and Lilly became partners in 1855; all their forging was by hand until 1865 when they installed a small steam hammer" (Baker 1973, 434). See Crooker & Lilly and David Crooker.

Lincoln, Robert**   Biddeford   -1855-

Tools Made: Brass Founders and Finishers

Remarks: He is listed in the 1855 *Maine Registry and Business Directory*.

Linscott, J. W.   Freedom   -1885-
Tools Made: Hoes

Linton, William**   Portland   1829
Tools Made: Gunsmith
Remarks: Demeritt (1973, 164).

Lithgow, William**   Thomaston, Brunswick, Fort Halifax (now Winslow)   1739-1754
Tools Made: Gunsmith
Remarks: Demeritt (1973, 11, 164) notes he moved from Thomaston to Brunswick in 1742 and to Fort Halifax in 1754.

Littlefield, Charles E.**   Carver's Harbor   1882
Tools Made: Tool for Dressing Mortises
Remarks: He has Pat. No. 128,050 for a tool for dressing mortises (a plane-like device), Jun. 18, 1882 (Tom Lamond, personal communications).

Littlefield, John B.**   Lewiston   1873-1892
Tools Made: Gunsmith
Remarks: Demeritt (1973, 11, 164).

Livingston Mfg. Co.   Rockland   1893-
Tools Made: Coach-making Tools, Hammers, and Stone-working Tools
Remarks: The Livingston Mfg. Co. used the marks: **LIVINGSTON | MFG. CO. | ROCKLAND.ME.** and **LIVINGSTON MFG. CO. | ROCKLAND ME.** The Livingston Co. was purchased by the Bicknell Mfg. Co. in the early 20th century.

Livingston, James**   Calais   -1881-
Tools Made: Shipsmith
Remarks: He is listed in the 1881 *Maine Business Directory*.

Lockhart, George**   Portland   -1869-
Tools Made: Shipsmith
Remarks: He is listed in the 1869 *Maine Business Directory* as located at 329 Commercial St.

Lombard, Allen   Augusta   -1855-
Tools Made: Farm Tools

London, Jonathan   Bridgewater   -1855-1856-
Tools Made: Edge Tools
Remarks: He is listed in the 1855 and 1856 *Maine Business Directories* as an edge toolmaker.

Long, Malcolm W.**   Bangor, Augusta   1855-1875-   (d. 1912)
Tools Made: Gunsmith
Remarks: Demeritt (1973, 164) notes he was first part of Graves & Long in Bangor, and then moved to Augusta in 1871, where he worked by himself. Finally, he moved to Harrisburg, PA, in the1880s

Lord & Graves*   West Waterville   1850-1858
Tools Made: Axes

Lord, D. P.**   Denmark, Oakland   -1860-1879-
Tools Made: Axes
Remarks: He is listed in Denmark in the 1879 *Maine Business Directory*. Klenman (1990) lists Daniel P. Lord as one of the important ax-makers on Emerson Stream in Oakland, Maine, working around 1860. This may be the same ax-maker as the listing below for Daniel B. Lord, if his middle initial was misread at some point. He also could be a relative or of no relation. Ax-makers and blacksmiths didn't always stay in the same location, and the exact relationship between the many Maine ax-makers from a particular family clan remain a mystery pending further research.

Lord, Daniel B.   Waterville   1850-1862
Tools Made: Axes, Farm Tools, and Hoes
Remarks: D. B. Lord is listed in the 1855 *Maine Registry and Business Directory* as located in West Waterville. He sold his business to Hubbard & Blake in 1862.

Lord, Horace**   Levant   -1867-
Tools Made: Edge Tools
Remarks: He is listed in the 1867 *Maine Business Directory*.

Lord, W. J.*   Sedgwick
Tools Made: Axes

Love & Thayer**   Augusta   1873-1876
Tools Made: Gunsmith and Locksmith
Remarks: Demeritt (1973, 164) notes this is Robert C. Love.

Lovejoy Brothers*   Chesterville   1878-1924
Tools Made: Axes
Remarks: This could have been a partnership including Leonard Lovejoy.

Lovejoy, Colins**   Chesterville, North Chesterville   -1855-1879-
Tools Made: Timber Framing Chisels and Axes
Remarks: C. Lovejoy of Chesterville is listed in the 1855 and 1856 *Maine Business Directories* as an edge toolmaker. The 1879 *Maine Business Directory* lists Colins Lovejoy of N. Chesterville and Chesterville as an ax-maker. This photograph is of an ax owned by Philip McKinney that is marked: **C. LOVEJOY | CHESTERVILLE | ???? | CAST STEEL**. A 2" chisel with this same signature has been reported by a collector in Farmington, ME. That chisel has a fourth line that says **WARRANTED**. The line above cast steel

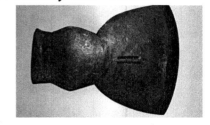

is not legible on the ax. The Davistown Museum has a timber framing chisel (ID# 91501T1, courtesy of Dana Philippi, 9/15/01) signed **C. LOVEJOY | CHESTERVILLE** in the MIV collection.

Lovejoy, Hubbard   Wayne, Auburn   1830-68   (b. 1807)
Tools Made: Wood Planes
Remarks: Lovejoy's planes had the imprint of an eagle with his name **H.LOVEJOY | WAYNE**. He first lived in Wayne and then moved to Auburn. It was in one of those towns that he worked with William Burgess ca. 1848-50 making doors, sashes, and blinds. He is believed to have otherwise worked as a builder (Nelson 1999). Pollak (2001, 259) mentions that the imprint has been found on a 28" jointer and a screw-arm plow, c. 1840-50. The beech jointer mentioned by Pollak is in the Bob Jones collection.

Lovejoy, Leonard R.   Chesterville   -1869-1878-
Tools Made: Axes
Remarks: His town was also reported as Chesterfield, which it is assumed is an error. He is listed in the 1869 *Maine Business Directory*.

Lovejoy, R.**
Tools Made: Axes
Remarks: Raymond and Michael Strout of Bar Harbor have recently discovered a broad ax signed **R. Lovejoy**, probably the Leonard R. Lovejoy listed above.

Low & Blunt**   Waterville   -1855-
Tools Made: Agricultural Implements
Remarks: They are listed in the 1855 *Maine Registry and Business Directory*.

Low, G. W.   Burlington   -1869-1871-
Tools Made: Axes
Remarks: He is listed in the 1869 *Maine Business Directory*.

Lowell & Senter   Portland   -1836-1870
Tools Made: Scientific Instruments
Remarks: Abner Lowell and William Senter's products included watches, jewelry, and nautical and survey instruments.

Lowell, Abner   Portland   1834-1877   (b. 12 Jan. 1812, d. 26 Feb. 1883)
Tools Made: Scientific Instruments
Remarks: Lowell worked as a watchmaker and jeweler before and after he was part of Lowell & Senter. It is not known if he made any instruments during those times by himself.

Luce, James   Troy   -1899-1900-
Tools Made: Farm Tools

Lucy, D. E.**   Houlton   1873-1878
Tools Made: Gunsmith and Jeweler
Remarks: Demeritt (1973, 164).

Lycett, James*   Springfield   1867-1881
Tools Made: Axes

Lyman, Enoch   Sullivan   -1855-1856-
Tools Made: Edge Tools
Remarks: He is listed in the 1855 and 1856 *Maine Business Directories*.

Lyon, Almon H.   Sidney   -1885-
Tools Made: Farm Tools, including Harrows

Mace, Albion K.   Smithfield   -1880-
Tools Made: Farm Tools, specifically Harrows and Cultivators

Macurda, A. H.   Lisbon   -1885-
Tools Made: Farm Tools

Maddocks, J.   Jackson   -1879-
Tools Made: Farm Tools

Maine Axe Co.**   Oakland

Tools Made: Axes
Remarks: They are listed by Lamond (2007b).

Mallett & Co.   Augusta   ca. 1881-
Tools Made: Screwdrivers
Remarks: He held an 1881 patent for a ratchet screwdriver.

Mallett, Charles H.**   Augusta   1881-1883
Tools Made: Screwdrivers
Remarks: See the Furbish biography for patent information. It is unknown if he was the part of Mallet & Co.

Mallett, James M.*   Warren   -1855-1882-
Tools Made: Axes, Ship Carpenters'
Tools, and Edge Tools
Remarks: He may have been the maker
of a mast shave (ID# 72801T1) marked
**MALLET CAST STEEL | WARRANTED
WARREN ME** in the Davistown Museum's
MIV collection (photo). He is listed in the 1855, 1856, 1867, 1869, 1881, and 1882
Maine Business Directories. Yeaton (2000) reports a James Mallet who worked as an ax-maker in Rockland prior to 1874 and up to 1892.

Mallet, John H.**   Rockland   -1855-1856-
Tools Made: Edge Tools, Shipsmith
Remarks: He is listed in the 1855 *Maine Business Directory.* The 1856 and 1881
Maine Business Directories list him as John L. Mallet. In 1881 he was located on Sea St.

Mann, Lewis M. & Son   West Paris   ca. 1900-1940-
Tools Made: Household Tools (Clothespins)

Mansfield, Edward   Orono   -1856-1906-
Tools Made: Axes, Edge Tools, and Other
Tools, including Cant Dogs and Peaveys
Remarks: Mansfield worked under his own
name in the 1870s in Orono but had moved to
Bangor and changed the name to E. Mansfield &
Co. by 1906. According to Yeaton (2000), this
company worked up to 1916. Tools were marked

E. MANSFIELD | ORONO, ME. He is listed in Orono in the 1856, 1869, and 1879 *Maine
Business Directories.* The Davistown Museum has a peavey (ID# 61204T1) in the IR

collection that is marked **E. MANSFIELD & Co | SNOW & NEALLEY CO | BANGOR, MAINE** (photo). Numerous peaveys with this signature have been sold by the Liberty Tool Co. suggesting a long time association of these two companies.

Marden, I. W.   Palermo   -1869-
Tools Made: Rakes and Planes
Remarks: A 1 ½" rounding plane signed **I. W. Marden** has been discovered by Dana Phillipi in Weeks Mills suggesting the probability that I. W. Marden and family of Palermo made more than just rakes.

Marden, J. A.   Veazie   ca. 1872
Tools Made: Marking Gauges
Remarks: The mark **J.A. MARDEN | PAT.APR.16 1872** has been found on marking gauges. Marden was issued the patent, but whether or not he was the maker is unknown. Pollak (2001, 266) states that this is possibly John A. Marden (b. 1823 in Palermo, ME, d. July 3, 1887 in Chicopee, MA) who was joiner and mill man in Bangor, ME, thru 1856. From 1860-1872, he was in Veazie, ME, and worked as a house carpenter. J. A. Marden was issued Patent No. 125, 823 on April 16, 1872 for a marking gauge. In 1873, John moved to South Hadley Falls, MA, then to Holyoke, MA, where he was a house carpenter and millwright. Example: 9 ½" beech 1/16" dado. See Bacheller (2000).

Mark, G & G**   Portland   -1850-1860-
Tools Made: Gunsmith and Cutler
Remarks: Demeritt (1973, 164) notes this was Gabriel Mark.

Mark, Godfrey**   Portland   -1850-1860-
Tools Made: Gunsmith
Remarks: Demeritt (1973, 164).

Marquis, Abel**   Fort Kent   -1881-
Tools Made: Axes
Remarks: He is listed in the 1881 *Maine Business Directory*.

Marr, J. F.   Lewiston   -1869-
Tools Made: Brushes (Brooms)

Marriner, D.**   BK   David
Tools Made: Planes
Remarks: Pollak (2001, 267) describes a 12" beech ships' rabbet plane in the Bob Jones collection marked **D.MARRINER | BK.Me.** Pollak suggests that this could refer to

Brunswick, Berwick, or some other coastal Maine town. Additional information on this mark is welcomed.

Marsh Axe & Tool Co.**   Oakland   -1925-
Tools Made: Axes
Remarks: He is listed in *American Axes* by Henry J. Kauffman (1972).

Marsh & Sons*   Oakland   1885-1926
Tools Made: Axes
Remarks: See the illustration of a Marsh & Sons label on page 26 of Klenman's (1990) *Axe Makers of North America*. Lamond (2007b) lists the working dates of 1885-1926.

Marshall, Joel**   Buxton   -1807-
Tools Made: Edge Tools
Remarks: He is listed in *American Axes* by Henry J. Kauffman (1972).

Marston, William P.   Bath   -1855-
Tools Made: Farm Tools

Martin & Tebbets**   Portland   -1821-
Tools Made: Iron and Steel
Remarks: This company was selling iron and steel as indicated by the reproduced advertisement from the *Independent Statesman* of Nov. 16, 1821.

Martin, James**   Thomaston   1758-1759
Tools Made: Gunsmith
Remarks: Demeritt (1973, 164).

Mason, George   Sidney   -1885-
Tools Made: Farm Tools

Mason, T. J.   Harmony   -1871-
Tools Made: Hoes (Horse Hoes)

Masterman, D. S.**   Weld   1877
Tools Made: Gunsmith
Remarks: Demeritt (1973, 164).

Mathews & Hubbard*   West Waterville   1854-1858
Tools Made: Axes

Mathews, Albert H.**   Augusta   1867
Tools Made: Gunsmith
Remarks: Demeritt (1973, 164) notes he worked with L. M. Leland.

Mathews, Isaac**   Thomaston   -1869-
Tools Made: Shipsmith
Remarks: He is listed in the 1869 *Maine Business Directory*.

Mathews, Walter**   Waldoboro   1856
Tools Made: Shipsmith
Remarks: He is listed in the 1856 *Maine Business Directory*.

Matthews, O. D.**   Thomaston   -1881-1882-
Tools Made: Shipsmith
Remarks: He is listed in the 1881 and 1882 *Maine Business Directories*.

Maxwell & Jameson   Topsham   1836-1873
Tools Made: Blacksmith, Axes
Remarks: This business is listed in the 1867 and 1869 *Maine Business Directories*. It was a joint venture of Samuel Jameson and James Maxwell.

Maxwell, Alexander   Milo   -1867-1871-
Tools Made: Axes and Shovels
Remarks: He is listed in the 1867 and 1869 *Maine Business Directories*.

Maxwell, Eben   Cape Elizabeth   -1869-
Tools Made: Plows

Maxwell, James**   Topsham   1836-1873
Tools Made: Blacksmith
Remarks: He worked with Samuel Jameson in the business Maxwell & Jameson.

Mayberry & Spurr   Otisfield   -1869-1871-
Tools Made: Edge Tools
Remarks: They are listed in the 1869 *Maine Business Directory*.

Mayberry, Richard**   Windham, Casco   1755-1807   (d.1807)
Tools Made: Blacksmith
Remarks: Richard was the son of William Mayberry. Following the American Revolution, he moved from Windham to Raymond (Casco). He was killed at age 72 in 1807 by a falling tree as he cleared land.

Mayberry, William**   Windham   1740-1765

Tools Made: Blacksmith

Remarks: Mayberry emigrated from Ballemmoney, Ireland, to Marblehead, MA, in about 1730. He settled in Windham in 1740, being the second to settle in the town. By 1850, he was reported to have a garrison house and 15 acres of cleared land. "He was by trade a blacksmith and brought with him the tools of his trade. A family tradition asserts that shortly after he came to Windham, for want of better accommodations, he set up his forge under the spreading branches of a gigantic oak tree near his dwelling, and placed his anvil on a convenient stump, prepared to exercise his old time handicraft, and that the Indians were his first customers" (Samuel Thomas Dole, Frederick Howard Dole and the Windham Historical Society).

Mayo & Sons   Foxcroft   -1885-

Tools Made: Farm Tools and Plows, including Cultivators

Remarks: See Mayo, J. B.

Mayo, Albion W.**   Portland   -1869-1882-

Tools Made: Shipsmith

Remarks: He is listed in the 1869 *Maine Business Directory*. The 1881 Business Directory lists him on Railroad wharf, and the 1882 Business Directory lists him on Railroad St.

Mayo, F. J.   Corinth   -1867-1871-

Tools Made: Axes

Remarks: Yeaton (2000) lists a Frank Mayo who worked in Corinth from 1869-1901. It is unknown if he was the F. J. Mayo listed in the 1867 and 1869 *Maine Business Directories* as an edge toolmaker in East Corinth and Corinth.

Mayo, H. M.**   Monmouth   -1879-1881-

Tools Made: Knives

Remarks: He is listed in the 1879 and 1881 *Maine Business Directories*.

Mayo, J. B.   Foxcroft   -1880-

Tools Made: Farm Tools, Hoes and Plows, including Cultivators and Horse Hoes

Remarks: One directory has listings for a J. G. Mayo (by himself) and a partnership of J. B. & J. G. Mayo, as well as J. B. Mayo. The pair may have been part of Mayo & Sons.

Mayo, William F.   Kenduskeag   -1862-1871-

Tools Made: Axes and Edge Tools

Remarks: Yeaton (2000) cites his working dates as being 1867-1887. He is listed in the 1867 and 1869 *Maine Business Directories*.

McCausland, A. W.   Gardiner   -1869-
Tools Made: Brushes, including Brooms

McDonald, William**   Sherman   -1879-
Tools Made: Axes
Remarks: He is listed in the 1879 *Maine Business Directory*.

McFarland, James**   Brunswick   1790-1797
Tools Made: Blacksmith
Remarks: McFarland apprenticed with Colonel William Stanwood. In 1790, McFarland took over the business. He continued until 1797, when he moved away and the shop was torn down (Wheeler 1878, 578). James was the step-son of Samuel Stanwood, an uncle of Col. William Stanwood. His father was John McFarland, and his mother was Jane Lithgow McFarland Stanwood, who remarried after his father died.

McFarlin, Charles**   Gardiner   1867-1880
Tools Made: Gunsmith
Remarks: Demeritt (1973, 164).

McKeen, Frank W.   Denmark, Lovell   -1867-1880-   (d. 1882)
Tools Made: Blacksmith, Axes, Knives, and Edge Tools
Remarks: He is listed as an ax-maker in the 1867 and 1869 *Maine Business Directories*. He is listed in the 1879 *Maine Business Directory* in Lovell. David Crouse of Bangor, Maine, has provided the following information "Frank W. McKeen, son of Eliphalet and Sarah (Ela) McKeen, was born in 1836, probably in Stoneham, Maine. He married Martha Sanders (1841-1879) of Pittsfield, NH, about 1860. The McKeens had five children. In the 1870 census of Denmark, ME, he is listed as a blacksmith. In the census of 1880, he is a blacksmith in Lovell, ME, where he died on 11 February 1882"

Mckeller, J. F.**   Rockland   1873
Tools Made: Gunsmith
Remarks: Demeritt (1973, 164).

McKenney**   Bangor   -1920-
Tools Made: Handsaws
Remarks: He had patent 1349427 on 8/10/1920 for an adjustable handsaw. Source: Graham Stubbs.

McKenney**   Biddeford
Tools Made: Saws
Remarks: Steve Beauregard of Hinckley, Maine has a 12

½" long early hacksaw marked **McKENNEY BIDDEFORD** (photograph courtesy of Rick Floyd). We don't know if this was made by the Bangor McKenney or one of the gunsmith McKenneys in Biddeford or someone else.

McKenney & Co.**
McKenney & Bean**   Biddeford   1855-1870-
Tools Made: Gunsmith
Remarks: Demeritt (1973, 97-9, 164) notes that Henry H. McKenney worked under both these company names. His partner was Samuel E. Bean.

McKenney, J. F.**   Biddeford   1850s
Tools Made: Gunsmith
Remarks: Demeritt (1973, 164).

McKenney, James**   Orland   -1856-
Tools Made: Shipsmith
Remarks: He is listed in the 1856 *Maine Business Directory*.

McKinney, W. W.   North Anson   -1869-
Tools Made: Rakes (Horse Rakes)

McLellan, John   Casco   -1869-
Tools Made: Brushes (Brooms)

McQuarris, John**   Bath   -1882-
Tools Made: Shipsmith
Remarks: He is listed in the 1882 *Maine Business Directory* on Water St.

Mead, Lewis P. & Co.   Augusta   -1849-
Tools Made: Farm Tools

Mead, N. & B.   Castine   -1810-
Tools Made: Augers
Remarks: They advertised in a Castine paper, but it is not clear if they actually worked there. See Mead, Noah.

Mead, Noah   Castine   -1810-
Tools Made: Axes
Remarks: Noah Mead advertised in the same paper and same issue in which N. & B. Mead's ad appeared.

Mears, J. R.   ?   -1879-1885-
Tools Made: Rakes (Horse Rakes)

Mechanic Falls**   ?   ca. 1850
Tools Made: Planes
Remarks: Pollak (2001, 276) reports a 8 5/8" spar plane in the Bob Jones collection that is marked **MECHANIC FALLS. ME**. The maker is unknown.

Mechanics Tool Co.   ?   ca. 1890
Tools Made: Planes
Remarks: DATM (Nelson 1999) states that it is not known if the company was a maker or dealer of planes marked **MECHANICS | TOOL Co. | WARRANTED**. A Mechanics Tool Co. of Boston, MA, possibly a totally different company, made a machinist's surface gauge ca. 1910 (Nelson 1999, 532). Pollak (2001, 276) mentions three planes with this mark, one an 8 ¼" beech smoother in the Bob Jones collection.

Merchant, Robert**   Berwick   1720
Tools Made: Rules

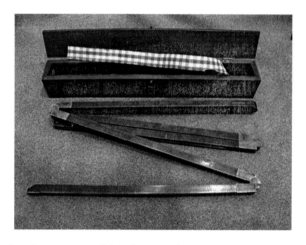

Remarks: There is no information about Robert Merchant in the DATM (Nelson 1999) or in any other publication, such as *The Chronicle*. That such a rule-maker lived and produced rules in Maine is verified by the fact that there is a signed rule that he made (ID# TBW1006) on display at the Davistown Museum in the MI collection. At publication of this registry, this beautifully constructed and elegant work of art, signed *Made by Robert Merchant for Noah Emery, Berwick, 1720* in script is the earliest known signed and dated tool made in the state of Maine and may very well be the earliest signed and dated measuring tool known to have been made in the United States. Robert Wheeler, who bought this tool at auction, speculates that Robert Merchant may be related to the many Merchants who lived at this time in and around Portsmouth, NH. Anyone who has additional information as to who Robert Merchant was, please contact the Davistown Museum. In addition to this photo, see the one on the front cover.

Merrick, George G.   Thomaston   ca. 1840-1860   (b.1814)
Tools Made: Wood Planes
Remarks: Merrick worked as a ships' joiner in Thomaston and made planes before he moved to Libertyville, IL, where he was not known to have made planes. According to DATM (Nelson 1999), he used two imprints on his planes: **G.G. MERRICK |**

THOMASTON and the more unusual **G G MERRICK | THOMASTON | M.E   M.E** with the name line curved and the two "M.E"s under either end of Thomaston. However, Pollak (2001, 277) mentions three imprints, all being rated as rare. The Bob Jones collection has a 9 3/8" cherry two piece sash plane with a fourth mark, with the **ME** under the curved **G.G.MERRICK** and above **THOMASTON**.

Merrill & Mead   Calais   -1869-
Tools Made: Brushes (Brooms)
Remarks: A W. R. Merrill made brooms in Manchester; any connection to this company, if there was one, is not known.

Merrill, A. C.**   Solon
Tools Made: Rules
Remarks: Phil Platt has reported a rule marked **A. C. MERRILL SOLON** that is believed to have been made in Maine.

Merrill, F. M.**   Freeport   -1840-
Tools Made: Edge Tools
Remarks: There is a 1 ¼ inch framing chisel signed by F. M. Merrill in the collection of Malcolm McFarland.

Merrill, Freeman   South Paris   -1869-1900-
Tools Made: Farm Tools and Plows, including Cultivators

Merrill, George**   Searsport   -1855
Tools Made: Pump and Block Makers
Remarks: He is listed in the 1855 *Maine Registry and Business Directory*.

Merrill, Joshua*   Freeport   1879-1917
Tools Made: Axes

Merrill, Josiah P.**   Freeport   1872-1890-
Tools Made: Gunsmith and Machinist
Remarks: Demeritt (1973, 165).

Merrill, Leonard F.**   Woodford's Corner   1878-1889
Tools Made: Gunsmith
Remarks: Demeritt (1973, 165).

Merrill, Samuel   West Gardiner   -1871-1885-
Tools Made: Plows

Merrill, W. R.   Manchester   -1869-
  Tools Made: Brushes (Brooms)
  Remarks: See Merrill & Mead

Merritt**   Solon
  Tools Made: Measuring Tools
  Remarks: Phil Pratt has a two foot long board stick marked "Merritt" and "Solon."

Metcalf, Joseph*   Winthrop   (b. 1756 d. 1849)
  Tools Made: Planes
  Remarks: One of Maine's earliest planemakers, several of his tools are in the

collection of the Davistown Museum (ID# TBW1009, photo). This registry contains an extensive discussion of his role in the early years of toolmaking in the introductory chapter "Maine's earliest planemakers." Also see his toolmaker information and biography page on the Davistown Museum's website.

Metcalf, William**   Thomaston   -1855-
  Tools Made: Pump and Block Makers
  Remarks: He is listed in the 1855 *Maine Registry and Business Directory*.

Meyers, Asa   Dresden   -1869-1871-
  Tools Made: Edge Tools
  Remarks: His name is also spelled as Myers in some directories. Which spelling is correct is not known.

Michand, Cyrel**   Van Buren   -1860-1870-
  Tools Made: Gunsmith
  Remarks: Demeritt (1973, 165).

Michaud, Joseph O.**   Fort Kent   -1907-
  Tools Made: Braces
  Remarks: This brace was patented on October 29, 1907 by Joseph O. Michaud in Fort Kent, Maine. This is the first one known to exist according to *Patented American Braces* by Ronald W. Pearson (1994). Information and photo courtesy of Rick Floyd.

Miles, William   Newport   -1849-

Tools Made: Edge Tools

Miller   Oakland
Tools Made: Axes
Remarks: His mark: **MILLER | OAKLAND, ME** has also been reported without the "**ME**".

Milliken, L. B.   Saco   -1899-1900-
Tools Made: Farm Tools

Mitchell, G.   Cambridge   -1879-1880-
Tools Made: Farm Tools

Monaghan, H.**   Whitneyville   -1855-
Tools Made: Wheelwright
Remarks: He is listed in the 1855 *Maine Registry and Business Directory.*

Monaghan, Stephen**   Ellsworth   -1855-
Tools Made: Wheelwright
Remarks: He is listed in the 1855 *Maine Registry and Business Directory.*

Monroe, Charles H.**   Searsport   1900
Tools Made: Sprocket Wrenches
Remarks: Patent #643,520 (Feb. 13, 1900) for a sprocket wrench was issued to Charles H. Monroe of Searsport, ME. We do not know if he made any wrenches. See the listing for the Sprocket Wrench Co. for a photograph and more information on this patent.

Mood, Samuel**   Augusta   1867
Tools Made: Gunsmith
Remarks: Demeritt (1973, 165).

Moody, B. C.**   Kingfield   -1855-
Tools Made: Agricultural Implements
Remarks: He is listed in the 1855 *Maine Registry and Business Directory.*

Moody, Samuel**   Brunswick   -1827-
Tools Made: Blacksmith
Remarks: A copy of Samuel's accounting day book is contained within the Special Collections at the Fogler Library, University of Maine. The day book gives a clear picture of the daily work of a nineteenth century blacksmith prior to the Civil War.

Moore & Cilley**   -1840-1850-
Tools Made: Planes
Remarks: Pollak (2001, 285) lists an example of a 14 1/2" beech jack with 19c chamfers marked **MOORE & CILLEY** in an arch, found in Maine.

Moor, B. L.*   Benton   -1874-1897
Tools Made: Axes

Moore, Hollis**   Lovell   1877
Tools Made: Gunsmith
Remarks: Demeritt (1973, 165).

Moore, J. C.**   Monson   1879
Tools Made: Gunsmith
Remarks: Demeritt (1973, 165).

Moore, William   Minot   -1849-
Tools Made: Edge Tools

Morgan, Isaac N.   Sedgwick   -1869-1871-
Tools Made: Edge Tools
Remarks: He is listed in the 1869 *Maine Business Directory*.

Morgan, Patrick**   Portland   -1882-
Tools Made: Shipsmith
Remarks: He is listed in the 1882 *Maine Business Directory* at 285 Comercial St.

Moriarty, Mathew   Bangor   ca. 1872-1906   (d.1911)
Tools Made: Coopers' Tools and Wood Planes
Remarks: Moriarty worked as a cooper making cisterns and coopers' tools to order, as noted in an 1873 directory. He arrived in Bangor, ME, sometime in the 1860s and operated a cooperage for 40 years, retiring in 1906 (Pollak 2001, 286). He held a 16 July 1872 patent for an adjustable howell. He used the mark: **M.MORIARTY | BANGOR.Me.** Pollak (2001) states that he may have apprenticed to Samuel Doyen in 1867. One of his yellow birch sun planes sold at a Brown Auction, October 29, 2005.

Morrill, Benjamin   Bangor   -1832-1851-   (b.1789, d.1862)
Tools Made: Wood Planes
Remarks: Son of a carriage maker, Benjamin Morrill was both a planemaker and joiner working in Bangor. His work as a planesman is first noted in the Dec. 4, 1832 minutes of the Bangor Mechanic's Association. He also served in the state legislature and

was a captain in the state militia. Pollak (2001, 287) describes his marks as **B.MORRILL |
BANGOR** or **B·MORRILL**. The Davistown Museum
has several planes by Morrill (ID# 032203T11
[photo,] 51100T14, 81101T1, and 61601T3) in the
MIII and MIV collections. The Bob Jones collection
contains one of his planes using the mark without the
place name.

Morrill, D. C.**
    Tools Made: Planes
    Remarks: Pollak (2001, 287) lists a 30" joiner marked **D.C. MORRILL** found in
Maine. (See Benjamin Morrill.)

Morrill, John J.   Hartland   -1849-1885-
    Tools Made: Axes and Edge Tools
    Remarks: He is listed in the 1855, 1856, 1869, 1879, and 1881 *Maine Business
Directories*.

Morrill, Joshua**   Cumberland   -1855-1856-
    Tools Made: Edge Tools
    Remarks: He is listed in the 1855 and 1856 *Maine Business Directories* as an edge
toolmaker.

Morrill, Levi   West Cumberland   -1871-
    Tools Made: Rakes (Horse Rakes)

Morrill, S.   ?   ca. 1800
    Tools Made: Planes
    Remarks: DATM (Nelson 1999) and Pollak (2001) mention planes with the mark
S.MORRILL. The Bob Jones collection contains a rounding plane with this signature that
was found in Maine.

Morris, Charles*   Newfield   1877-1900
    Tools Made: Axes
    Remarks: The 1881 *Maine Business Directory* lists a C. R. Morris and the 1882 *Maine
Business Directory* lists C. W. Morris. It is unknown if the Charles Morris of Newfield
listed by Yeaton (2000), C. R. Morris, and C. W. Morris are the same or different people.

Morrison & Joy**   Ellsworth
    Tools Made: Handsaws
    Remarks: A nine point cross cut saw was found in Bar Harbor by Michael Strout; the

saw is clearly marked with an etched signature **MORRISON & JOY | ELLSWORTH MAINE** in the same location on the saw blade where Disston and other companies usually etch their imprint. The saw has a Disston style brass with three brass nuts in a traditional style carved handle. David Williams of Ellsworth, Maine, notes that Morrison & Joy was a store in Ellsworth selling hardware and groceries that was founded by Austin Joy and C. W. Morrison in 1879. It operated into the 1930s.

Morrison, Duncan**   Portland   1867
Tools Made: Gunsmith
Remarks: Demeritt (1973, 165).

Morse, Charles R.   Newfield   -1867-1871-
Tools Made: Axes
Remarks: He is listed in the 1867 and 1869 *Maine Business Directories*.

Morse, J. B.**   Dixmont   -1855-1856-
Tools Made: Edge Tools
Remarks: He is listed in the 1855 and 1856 *Maine Business Directories* as an edge toolmaker.

Morse, J. H.**   Mt. Vernon   -1867-
Tools Made: Edge Tools
Remarks: He is listed in the 1867 *Maine Business Directory*.

Morse, Nathan   Smithfield   -1869-
Tools Made: Farm Tools (Cultivators)

Morse, S. C.*   Milford   1875-1885
Tools Made: Axes

Morton, H. A.   Farmington   -1880-1885-
Tools Made: Plows

Morton, P.   Hallowell   -1849-
Tools Made: Farm Tools
Remarks: First initial might actually be "D".

Morton, R. L.   Farmington   -1869-1871-
Tools Made: Farm Tools
Remarks: It is unknown if he is related to H. A. Morton.

Moses, William V. and Oliver**   Bath   -1850-
Tools Made: Shipsmith
Remarks: Information from Baker (1973) *Maritime History of Bath, Maine.*

Mosher, C. A. & Co.**   Auburn   -1881-
Tools Made: Edge Tools
Remarks: This company is listed on Main St. in the 1881 *Maine Business Directory.*

Mosher, George**   Presque Isle   1856-1883
Tools Made: Gunsmith and Blacksmith
Remarks: Demeritt (1973, 165).

Moulton, Josh**   Scarborough   -1832-
Tools Made: Blacksmith
Remarks: He is listed in *American Axes* by Henry J. Kauffman (1972).

Moulton, George**   Bath   1838-1842
Tools Made: Blacksmith
Remarks: After 1842, he went into business to build steam engines and boilers with John H. Williams. Baker (1973), notes that the first steam boiler built in Bath was built by Moulton and Williams. After 1846, Moulton bought out Williams and continued in operation as George Moulton & Company and then founded the Commercial Street Ironworks in 1854, making "all kinds of machinery, steam engines, boilers, ships water tanks, and the various articles of ships work such as wooden and iron capstans, steering wheels of wood or iron, planking and jack screws--all the different kinds of ships pumping apparatus, also iron fences of beautiful patterns" (Baker 1973, 435). The career of George Moulton clearly illustrates the transition from the age of the hand work by traditional shipbuilders, with their adzes and pitsaws, to the age of steam-powered woodworking machinery, which, after 1850, was used for much of the woodworking done by the sawyer in Bath's shipyards.

Mowry, Bradley R.   Union   ca. 1820-1860   (b. 1796)
Tools Made: Adzes and Edge Tools
Remarks: He is listed in the 1855 and 1856 *Maine Business Directories.* All tools marked **B.R.MOWRY** that have been reported are coopers' adzes. Rick Floyd has found a coopers' shave (photo) marked **MOWRY** with two starbursts in Damariscotta, Maine. It is unknown if this was Bradley Mowry, his son, M. H. Mowry, or someone else. It is known that Bradley had a son, Augustus, born in 1832, who worked with him during the years 1850-60, but it is not known if he continued after his father stopped. Sibley (1851) notes Mowry's location as at the

"Middle Bridge." See the quote under Vaughan & Pardoe. The Davistown Museum has a drawshave marked **B.R.MOWRY** in the MIV collection (ID# 31908T33).

Mowry, M. H.   Union   -1862-
Tools Made: Edge Tools
Remarks: Perhaps related to Bradley Mowry.

Munch, Ira F.**   Sumner   1873
Tools Made: Tool Holder
Remarks: He has Pat.No. 136,450, for an improvement in tool holder, Mar. 4, 1873. One half is assigned to David Torrey of Deering, ME (Tom Lamond, Personal communication).

Munroe, M. W.   West Troy
Tools Made: Edge Tools, specifically Froes
Remarks: Tools were marked **M.W.MUNROE | WEST TROY.ME.** Tom Ware of Troy contacted the Davistown Museum to report that he purchased a hand-forged M. W. Munroe butcher saw (20" long, and 15" long, 1" wide blade) with a riveted and peened handle.

Murch & Chapman**   Ellsworth   -1855-
Tools Made: Pump and Block Makers
Remarks: They are listed in the 1855 *Maine Registry and Business Directory*.

Murch, L. C. & Son**   Belfast   -1855-
Tools Made: Pump and Block Makers
Remarks: They are listed in the 1855 *Maine Registry and Business Directory*.

Murray, Walter David**   Portland
Tools Made: Spokeshaves
Remarks: This photo is of a drawshave marked

MURRAY NO 3 PAT JULY 30 '01. Information about Walter David Murray can be found in Thomas Lamond's (1997, 193-5) *Manufactured and patented spokeshaves & similar tools: Identification of the artifacts and profiles of the makers and patentees*. An example of a No. 2 model has also been identified. Mr. Lamond believes that Murray was the only individual from Maine to patent a true spokeshave.

Myars, F. A.**   Addison   -1855-
Tools Made: Pump and Block Makers
Remarks: He is listed in the 1855 *Maine Registry and Business Directory*.

Myers, Asa**   Dresden   -1867-1869-
Tools Made: Axes
Remarks: He is listed in the 1867 and 1869 *Maine Business Directories.*

Nason & Eastman**   Lewiston   1896-1900
Tools Made: Gunsmith
Remarks: Demeritt (1973, 165).

Nason, Charles F.**   Lewiston   -1860-1900-   (d. 1909)
Tools Made: Gunsmith
Remarks: Demeritt (1973, 85-6, 165) notes he worked by himself and starting in 1896 with Ivan F. Eastman as part of Nason & Eastman.

Nason, Elbridge G., Jr.**   Auburn   1864 - 1898
Tools Made: Gunsmith
Remarks: Demeritt (1973, 165) notes that in 1898 he worked with C. F. Nason.

Nason, G. W.**   Bath   1876-1877
Tools Made: Gunsmith
Remarks: Demeritt (1973, 165).

National Line and Cordage Machine Company**   Norway   circa 1880-1891-
Tools Made: Machines for making Cord and Rope
Remarks: Thomas Brown Dooley of Chelsea, MA received Letters Patent for "Improvement in machines for making cord and rope" in 1891, with the comment "This patent was taken out in the name of Thomas Wood Norman in Norway." Norman was treasurer of the Maine company. Both he and Dooley held shares in the company. The Davistown Museum has an information file on this company and Thomas Norman's patented line and cordage machine in the Center for the Study of Early Tools library.

Neal, Andrew**   Lincoln   1865
Tools Made: Gunsmith
Remarks: Demeritt (1973, 165).

Neal, Elijah**   York Co.
Tools Made: Blacksmith
Remarks: He is listed in *American Axes* by Henry J. Kauffman (1972).

Neal, George W.   Orono   -1879-
Tools Made: Edge Tools
Remarks: He is listed in the 1879 *Maine Business Directory.*

Neal, John H.** Bangor 1864-1878 (d. 1917)
Tools Made: Gunsmith
Remarks: Demeritt (1973, 74, 165) notes he was partnered with Charles V. Ramsdell in Ramsdell & Neal.

Neal, William** Bangor 1845-1853 (d. 1853)
Tools Made: Gunsmith
Remarks: Demeritt (1973, 63-8, 165).

Neal, William H. Saco -1869-
Tools Made: Brushes

Nealley, James** Winterport 1810- (b.1793)
Tools Made: Blacksmith
Remarks: He was born in Winterport in 1793 and learned the trade of blacksmith at which he worked, in connection with farming, throughout his life (*Biographical Review 1897,* 392).

Neil, E. H. Skowhegan -1849-
Tools Made: Farm Tools

Nelson, J. P. & Son Corinna -1879-1880-
Tools Made: Farm Tools

New England Axe Company** Oakland -1905-
Tools Made: Axes
Remarks: Klenman (1990), lists this company as one of the many Oakland, Maine, ax-makers located along Emerson Stream. This company is not listed in DATM (Nelson 1999), possibly because it was not incorporated until after 1900, the cutoff point for DATM listings.

Newbegin, E. G.** Old Town -1879-
Tools Made: Edge Tools
Remarks: He is listed in the 1879 *Maine Business Directory*. It is unknown if this the same person as the Newbegin in Milford.

Newbegin, Edward G. Milford -1856-
Tools Made: Blacksmith and Chisels
Remarks: Newbegin was a blacksmith who made tools and marked them **E.G. NEWBEGIN | MILFORD, ME.**

Newbegin, John   Gray   -1873-1881-
Tools Made: Axes
Remarks: He is listed in the 1879 and 1881 *Maine Business Directories*.

Newcomb, H. W.**   Eastport   1863-1868
Tools Made: Gunsmith
Remarks: Demeritt (1973, 165).

Newell, S. J.   Weeks Mills   -1871-
Tools Made: Farm Tools, including Rotary Harrows

Newhall, H. C.   Canaan   -1849-
Tools Made: Farm Tools

Newton, C. A.**   Dixfield   -1882-
Tools Made: Edge Tools
Remarks: He is listed in the 1882 *Maine Business Directory*.

Newton, E. M.**   Skowhegan   1861-1869
Tools Made: Gunsmith
Remarks: Demeritt (1973, 165).

Nicholas, James & Co. **   Calais   -1855-
Tools Made: Pump and Block Makers
Remarks: They are listed in the 1855 *Maine Registry and Business Directory*.

Nichols, David**   Berwick   1759-1784   (b. 1734 d. 1775)
Tools Made: Blacksmith, Axes, Broad Axes, Drawknives, Scythes, Hoes, Forks, and Plows
Remarks: One of Nichols's accounting books is in the University of Maine Fogler library special collections. The account lists the daily activities of an eighteenth century blacksmith, the tools and materials a blacksmith was expected to manufacture or mend, and the items and services bartered as payment for blacksmith services. He was a Quaker and married in 1758.

Nicholson, John**   Union   -1790-
Tools Made: Planes
Remarks: John Nicholson was one of southern Massachusetts' most famous and prolific planemakers. Late in life, he moved to Union, Maine (exact date unknown). No planes with a Nicholson signature marked "Union, Maine," have yet surfaced, but there remains the possibility that he made a few hand planes after his move to Maine.

168

Noble, John**   Brunswick   1825-1838
  Tools Made: Blacksmith

Nolin, G. & N.   Skowhegan   -1879-1899-
  Tools Made: Farm Tools, Scythes, and Sickles
  Remarks: It is not known if G. Nolin and N. Nolin worked together. However, they were both listed as scythe-makers in 1879 and 1885. G. Nolin held a 23 Feb. 1886 patent for a grass hook. He may have also made that tool, which was listed for sale in 1899. In 1901, a Nolin Mfg. Co. made scythes, grass hooks, hay knives, etc. in Skowhegan, but what involvement (if any) either Nolin had in this company is not known, as a "George Underwood" was listed as its president (Nelson, 1999). G & N Nolin is listed in the 1879 *Maine Business Directory*. G. & M. Nolin is listed in the 1881 and 1882 *Maine Business Directories* and possibly the M is a typo.

North Wayne Scythe Co.   Wayne, Fayette   1845-1900-
  Tools Made: Axes, Edge Tools, Farm Tools, and Scythes
  Remarks: According to DATM (Nelson 1999), the information for this company that began in Wayne and later operated in Fayette, is confusing and contradictory. "This name was reported from directories through 1900, but one reporter of such sources says they became the North Wayne Tool Co. without indicating when that occurred. One secondary source says they were 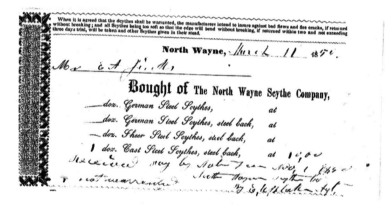 bought out by the Dunn Edge Tool Co. in 1865 and run by them with this name until 1887; another says they ceased business in 1870. An implied mark on a tool (**NORTH WAYNE CO. | J.F. TAYLOR, AGT.**) may have been made by this company or the North Wayne Tool Co. and some of the contradictory data may reflect confusion between these two companies." The Wayne Historical Society published *A History of The North Wayne Tool Co.* (Kallop 2003). It gives a chronology of the various names used by this company, which is listed in the 1855 *Maine Registry and Business Directory* as located in Fayette, in the 1856 *Maine Business Directory* as located in North Wayne, and in the 1867 and 1869 *Maine Business Directory* as located in Wayne. Figure of the bill is courtesy of Raymond Strout.

North Wayne Tool Co.   Hallowell   1879-1969

Tools Made: Axes, Farm Tools, Scythes, Lumbering tools, Corn Hooks, and Hay Knives

Remarks: Charles W. Tilden, Joseph E. Bodwell, and Williston Jennings operated this company in Hallowell, West Waterville, and Wayne, and then moved it to Oakland around 1900. They marked their tools: **NO. WAYNE TOOL CO | OAKLAND.ME.U.S.A.** and **NORTH WAYNE TOOL CO. | OAKLAND, MAINE** and used the brand name **LITTLE GIANT** (probably after 1900) (Kallop 2003). Dana Phillippi has this ax (photo) marked **NWTCO** and **3** in addition to the paper label.

Norton, Asa H.   Bangor   -1834-1856-

Tools Made: Rules, specifically Wantage Rods and Log Rules

Remarks: He is listed in 1855 directories as a rule maker; he also worked as a jointer or carpenter. "In 1861, the ME Legislature prescribed the use of Norton or Holland rules; it is not certain that Norton was still working then himself, so the prescription may be for some design or scaling characteristic vs. rules specifically made by Norton" (Nelson, 1999). The following marks were used on his rules: **ASA H. NORTON | MAKER BANGOR ME** and **ASA NORTON'S | LOG RULE | CORRECTED 1842**. Thank you to Rick Floyd for providing the photograph of this wantage rule.

Norton & Sprague**   Farmington   -1855-

Tools Made: Agricultural Implements

Remarks: They are listed in the 1855 *Maine Registry and Business Directory.*

Noyes, I. M.**   Machias   -1869-

Tools Made: Shipsmith

Remarks: He is listed in the 1869 *Maine Business Directory.*

Nute, Alexander**   Wiscasset   -1869-

Tools Made: Shipsmith

Remarks: He is listed in the 1869 *Maine Business Directory.*

Nutter, G. W.   Corinna   -1885-

Tools Made: Farm Tools, Hoes, and Plows, including Corn Shellers, Harrows, Cultivators, and Horse Hoes

Nutting, Ebenezer**   Falmouth (now Portland)   1772-1726
Tools Made: Gunsmith
Remarks: Demeritt (1973, 165).

Nutting, Mighill**   Portland   -1830-1850-
Tools Made: Gunsmith
Remarks: Demeritt (1973, 41-50, 165) notes he has several patents.

Nutting, Nathan   Otisfield   -1850-1867   (b.1804, d.1867)
Tools Made: Wood Planes
Remarks: The DATM (Nelson 1999) indicates that a Nathan Nutting, who is assumed to have been this planemaker's father, moved from Groton, MA, to Maine in 1795; any other relation to the Nuttings who made planes in MA is uncertain. Pollak (2001, 302-3) reports that Nathan Nutting started out as a journeyman carpenter in 1833 and continued to be listed as such in the 1855 and 1867 business directories, except that the 1850 census lists his occupation as a mechanic. His mark **N.NUTTING | OTISFIELD** has been reported on a large hollow plane in the Bob Jones collection.

O., T.**   Old Town   -1840-
Tools Made: Planes
Remarks: Pollak (2001, 308) states that this is possibly Thomas Otois (Otis) who worked as a sash-maker or joiner in Old Town, north of Bangor, ME. Examples, marked **T.O.**, include a number of planes found in an old tool chest, including a fruitwood plow, a beech double-bladed fixed sash plane, a beech and an applewood complex molder, and two beech coping planes, all 9 ½" with shallow round chamfers. One example is on a plane made by **T.Waterman | Waldoboro, ME** as an owner's imprint.

O'Connor, Michael W.**   Sanford   1881-1900
Tools Made: Gunsmith and Machinist
Remarks: Demeritt (1973, 165).

Orber, Joseph F.**   Mount Desert   1872
Tools Made: Shoemaker's Combination Tool
Remarks: He has Pat. No.133,383 for a shoemaker's combination tool (pinches, hammer, and awl), Nov. 26, 1872 (Tom Lamond, personal communication).

Orris, Jonathan**   Portland   Prior to 1690
Tools Made: Blacksmith

Osgood, J. K.**   Houlton   1875
Tools Made: Gunsmith

Remarks: Demeritt (1973, 165).

Otois, Thomas (see T. O.)

Ouellet** East Freeport -1820-
Tools Made: Planes
Remarks: Pollak (2001, 308) states that this person is believed to be Paul Ouellet from East Freeport, ME. An example, marked **E:FRE.ME.OUEL/LET**, is a 9" beech skew rabbet with heavy round chamfers of French Canadian influence.

Owen, A.** Trescott 1877
Tools Made: Gunsmith
Remarks: Demeritt (1973, 165).

Paine, B. Livermore Falls -1879-1885-
Tools Made: Plows
Remarks: The town East Livermore has also been cited in some years; the two towns are probably the same.

Paine, Freeman Standish -1869-1871-
Tools Made: Axes and Edge Tools
Remarks: He is listed in the 1869 *Maine Business Directory*.

Palmer, Hiram Whitefield -1871-
Tools Made: Rakes (Horse Rakes)

Palmer, John F. Whitefield -1885-
Tools Made: Rakes (Horse Rakes)

Palmer, R. L.** Farmingdale -1855-
Tools Made: Edge Tools
Remarks: He is listed in the 1855 *Maine Business Directory*.

Palmer, V. S. Kenduskeag -1879-1885-
Tools Made: Farm Tools, Plows, and Rakes (Harrows, Horse Rakes, etc.)

Paris Hill Mfg. Co. Paris -1879-1885-
Tools Made: Farm Tools

Park & Sons* West Waterville
Tools Made: Axes

Parker, F. J.**   Bucksport   -1869-
Tools Made: Shipsmith
Remarks: He is listed in the 1869 *Maine Business Directory*.

Parker, John M.**   Brunswick   1867-1880
Tools Made: Gunsmith
Remarks: Demeritt (1973, 165).

Parker, Melvin B.**   Waterville   1909
Tools Made: Gunsmith
Remarks: Demeritt (1973, 165).

Parks, Samuel H.   Orono, Waterville   1850-1873
Tools Made: Axes
Remarks: He also worked in Waterville. It is unknown if he worked in both towns at the same time or moved from one to the other.

Parris, A.**   Portland   -1800-1820-
Tools Made: Planes
Remarks: Pollak (2001, 312) states that this person is believed to be Alexander Parris who was a joiner and builder in Portland, ME. Examples, marked **A+PARRIS**, include a 10 7/16" complex molder, and a 10 1/16" complex molder with rosewood boxing, both birch with medium flat chamfers and the A wedge, and a 10 1/8" birch double-iron fixed sash with tight round chamfers and the B wedge.

Parsons, A.**   -1810-1820-
Tools Made: Planes
Remarks: Pollak (2001, 313) lists the following examples, marked **A.PARSONS**, a 31 1/2" beech jointer with heavy flat chamfers and a 9 1/2" beech round with an iron set at a "York" pitch with large round chamfers found in Maine.

Parsons, H. M.**   Eddington   -1855-
Tools Made: Wheelwright
Remarks: He is listed in the 1855 *Maine Registry and Business Directory*.

Parsons, John H.**   Augusta   1874-1899   (d. 1899)
Tools Made: Screwdrivers and Gunsmith
Remarks: Demeritt (1973) notes he was partnered with George Gay of Gay & Parsons and Charles Henry Safford of Safford & Parsons (1874-1875).

Parsons, Joshua** Minot -1855-
Tools Made: Forks
Remarks: He is listed in the 1855 *Maine Registry and Business Directory*.

Partridge, Frank R.** Augusta
Tools Made: Screwdrivers
Remarks: Partridge held patent 266642, 10/31/1882 for a reversible screwdriver. Information courtesy of C.D. Fales.

Passmore, Young & Taft Waterville 1849-1854
Tools Made: Scythes

Patten, George M. & Co. Bath -1869-
Tools Made: Farm Tools (Mowing Machines)

Patten, James T. & Co. ** Bath -1855-
Tools Made: Spikes
Remarks: They are listed in the 1855 *Maine Registry and Business Directory*.

Patten, John** Topsham -1750-
Tools Made: Blacksmith
Remarks: Patten came to Topsham in 1750. He was a farmer but had the trade of blacksmith and had a shop on his farm, where he employed a portion of his time, and performed the blacksmith work of the vicinity (Wheeler 1878, 610).

Payson* Hope 1847-1870
Tools Made: Axes
Remarks: An ax marked **PAYSON | SO. HOPE, ME** has been reported to the museum (photograph from Iron Horse Sculpture). It is unknown how these three Payson listings are related.

Payson & Howard** S. Hope
Tools Made: Socket Chisels
Remarks: James Hill of Warren, Maine has found a ½" socket chisel signed **Payson & Howard So Hope ME Warranted**. It is 18" long with a 4" handle and made of cast steel.

Payson, John & Son** Hope -1856-
Tools Made: Edge Tools
Remarks: This company is listed in the 1856 *Maine Business Directory* as an edge toolmaker.

Peabody, W.**   Orono   -1826-
Tools Made: Axes
Remarks: He is listed in *American Axes* by Henry J. Kauffman (1972).

Peabody, W.*   Orono   -1874-1876
Tools Made: Axes

Pearson, Henry Sleeper   Portland   -1823-1878-   (b. 23 May 1789, d. 30 Aug. 1878)
Tools Made: Scientific Instruments and Wantage Rods (?)
Remarks: Pearson's main occupation was as a watchmaker, but he also made mathematical instruments, which included survey compasses.

Peaslee, John T.   Alna   -1869-1879-
Tools Made: Edge Tools and Axes
Remarks: He is listed in the 1869 and 1879 *Maine Business Directories*.

Peavey Brothers*   Bangor   -1874-1879
Tools Made: Axes

Peavey Mfg. Co.*   Bangor   1857-
Tools Made: Axes, Logging, Pruners, Saws, Ice Scrapers, Shovels, Crow Bars, and Pry Bars
Remarks: Peavey Mfg. Co. is listed among Yeaton's (2000) ax-makers. From 1900 to 1918, the company was in Bangor, in Brewer from 1918 to 1923, and then in Oakland from 1927 to 1965. The DATM (Nelson, 1999) mentions it under Peavey Tool Co., which they apparently bought out in 1928. The Peavey Manufacturing Company is now located in Eddington, Maine. The company claims to have been in operation since 1857. A copy of a receipt from Bangor Edge Tool Co., which is also marked "Peavey Manufacturing Co.," is reproduced in the figure for the Bangor Edge Tool Co. listing. For more information, see the Davistown Museum website information file for the Peavey Mfg. Co.

Peavey Tool Co.   Bangor/Oakland   -1928
Tools Made: Peaveys
Remarks: Supposedly started by two grandsons of Joseph Peavey, creator of the "peavey" cant-dog around 1860, this possibly pre-1900 company continued to make that tool until around 1928, when it was taken over by a Peavey Mfg. Co. in Brewer, Maine. This is curious because the mark being used prior had been **THE PEAVEY MFG. CO. | OAKLAND, MAINE**. The names, relationships, and sequence of Peaveys who apparently made this tool are confusing. The Davistown Museum has a Peavey Mfg. Co., Oakland peavey (ID# 4106T9) in the IR collection.

Peavey, Andrew**   Liberty, South Montville   -1850-1875   (d. 1897)

Tools Made: Blacksmith and Gunsmith

Remarks: He is listed as the son of William, living in the same Liberty household in the 1850 census. Demeritt (1973, 99-103, 165) lists him as a gunsmith living in S. Montville from 1856-1875. He moved to Lawrence, MA, in 1876, where he continued working until 1897. He has an 1876 patent.

Peavey, C. A. & J. H.*   Bangor   1879-1896

Tools Made: Axes

Remarks: This is another variation of the many incarnations of the Peavey clan of timber-harvesting toolmakers. This company name is particularly intriguing because of the discovery of stationary and a catalog by Bar Harbor collector Michael Strout dated 1887 on the postmark. While the catalog is titled "Bangor Edge Tool Company," this company is a vendor of Peavey tools and has on its stationary the name C. A. Peavey. A photograph of the catalog envelope is reproduced under the Bangor Edge Tool Co. listing. The entire catalog is reproduced within the museum's Peavey Mfg. Co. information file (Brack, 2008b) and contains illustrations of the most important tools manufactured by the Peavey group of companies. It is unknown whether the above company made tools other than axes and was, therefore, one of several locations where their line of tools was produced or whether they only made axes at this location.

Peavey, Frank   Smithfield   -1879-

Tools Made: Rakes

Peavey, James Henry   1879-1900-

Tools Made: Other

Remarks: James Henry may have been associated with Peavey Tool Co. and/or possibly the Peavey Mfg. Co. He reportedly patented a cant-dog, an improved version of Joseph Peavey's tool. However, their relationship is not known, and his modifications may have come after 1900.

Peavey, Joseph   Stillwater   1858

Tools Made: Other

Remarks: Joseph, a blacksmith and inventor, was the original creator of the "peavey" cant-dog lumberman's tool. Other inventions that have been accredited to him, but may not necessarily have been his, include the hay press, clapboard machine, water wheel, shingle machine, hoist, wooden vise, and an un-spillable inkwell. See the Davistown Museum's website information file on lumbering in Maine for a description of the invention of the peavey.

Peavey, S. H.* Bangor -1874-1897
Tools Made: Axes

Peavey, Thomas H.** South Montville, Liberty -1850-1860- (d. 1907)
Tools Made: Gunsmith
Remarks: Demeritt (1973, 165) notes he had a patent.

Peavey, Thomas J.** Liberty, Montville -1850-1870- (d. 1895)
Tools Made: Blacksmith and Gunsmith
Remarks: Thomas is listed as the son of William and living in the same household as William in the 1850 census. By the 1860 census, he is listed as a blacksmith in Montville. Demeritt (1973, 165) gives his residence as South Montville.

Peavey, Warren* Burlington 1879-1891
Tools Made: Axes
Remarks: He is listed in the 1881 *Maine Business Directory* as Warren Peavy.

Peavey, William** Liberty -1850-
Tools Made: Blacksmith
Remarks: He is listed in the 1850 census.

Peavy, William** Burlington -1867-1882-
Tools Made: Edge Tools
Remarks: He is listed in the 1867 and 1882 *Maine Business Directories*.

Penney & Thurston** Mechanics Falls 1864
Tools Made: Wrenches
Remarks: The Davistown Museum had a patented Penney & Thurston wrench in the IR collection, but it was stolen in the summer of 2003. See the museum's Penney & Thurston information file for patent information and other photographs of Penney & Thurston wrenches and the maker's mark. The figure is of a 12" quick adjusting nut wrench owned by Herb Page. It is clearly stamped with the J. W. Penney Oct. 11, 1864 patent date and Mechanics Falls, Me. It has a rack and trigger/pawl adjustment mechanism on the back side of the shank.

Penny, Jonathan** Sedgwick -1856-
Tools Made: Shipsmith
Remarks: He is listed in the 1856 *Maine Business Directory*.

Perkins, George W. Portland -1855-
Tools Made: Farm Tools

Perkins, Luke   East Winthrop   -1838-1839-
Tools Made: Hoes

Perkins, Stephen**   Kennebunkport   -1855-
Tools Made: Pump and Block Makers
Remarks: He is listed in the 1855 *Maine Registry and Business Directory*.

Perry, B. C.**   Buckfield   -1855-
Tools Made: Edge Tools
Remarks: He is listed in the 1855 *Maine Business Directory* as an edge toolmaker. Note there are three B. C. Perrys.

Perry, B. C.**   Sherman, Island Falls   -1867-1882
Tools Made: Axes
Remarks: He is listed in Sherman in the 1867 *Maine Business Directory*. He is listed in Island Falls in the 1882 *Maine Business Directory*. It is unknown if this was Barnabus C. Perry or a relative.

Perry, Barnabus C.*   Island Falls   1883-1894
Tools Made: Axes

Perry, Henry**   Patten   -1856-
Tools Made: Edge Tools
Remarks: He is listed in the 1856 *Maine Business Directory*.

Perry, John*   Island Falls   1902-1917
Tools Made: Axes
Remarks: It is unknown if he was a son or other relative of Barnabus Perry.

Perry, O. H.**   Rockland   -1860-
Tools Made: Gunsmith
Remarks: Demeritt (1973, 165).

Peters, David**   Eastport   -1855-
Tools Made: Wheelwright
Remarks: He is listed in the 1855 *Maine Registry and Business Directory*.

Philbrick, Joseph   Greenfield   -1885-
Tools Made: Farm Tools

Philbrick, R. R.   Somerville   -1867-1871-
Tools Made: Axes, Chisels, and Shovels
Remarks: He is listed in the 1867 and 1869 *Maine Business Directories*.

Philbrook & Pyne**   Bangor   -1880-
Tools Made: Gunsmith
Remarks: Demeritt (1973, 165) notes this was Francis J. Philbrook.

Philbrook, Francis J.**   Bangor   1871-1880-
Tools Made: Gunsmith
Remarks: Demeritt (1973, 165) notes he was partnered with Charles G. Staples of Staples & Philbrook in 1874 and part of Philbrook & Pyne in the 1880s.

Phillips & Archer   Bangor   -1869-
Tools Made: Brushes (Brooms)

Phillips & Bellmore**   Skowhegan   1896
Tools Made: Gunsmith
Remarks: Demeritt (1973, 165).

Phillips, Isaac A.**   Kingfield   -1855-
Tools Made: Edge Tools
Remarks: He is listed in the 1855 *Maine Business Directory*.

Phillips, Russell   Gardiner   1867-1870
Tools Made: Marking Gauges and Metal Planes
Remarks: Phillips held quite a few patents, three for planes and one for a marking gauge. His 13 Aug. 1867, 30 Aug. 1870, and 24 Oct. 1871 patents were used by Babson & Repellier, the Boston Tool Co., and C.C. Harlow to make metal planes. It is unknown who made marking gauges with his 15 Jan. 1867 patent, although it is implied that his name and city/state appear on the gauge. The 1867 patents were issued to him in Maine and the others in Boston, MA. DATM (Nelson 1999) notes that, "a few presentation models of the 1867 patent plane marked by Babson & Repellier are also marked **The Phillips Plow Plane Co. | Boston | Mass.**; no such company is known to have ever actually existed and the rationale for such a marking is unknown." According to Smith (1981, 93 - 6), Phillips moved to Boston before 1870. His Jan. 1867 patent was for a double marking gauge.

Phillips, William   Ashland   ca. 1881
Tools Made: Hammers and Pliers

Remarks: He held a 9 Aug. 1881 patent for a combination hammer and pliers tool but may not have been the maker. He may not have been the maker.

Phinney, Isaac E.   Stockton   -1862-
Tools Made: Edge Tools and Farrier Tools

Phipps, James**   Pemaquid, Bristol, Woolwich   1625-1654
Tools Made: Blacksmith and Gunsmith
Remarks: Apprenticed to John Brown for eight years, beginning in 1625. He eventually purchased land from John Brown, which was called Phipp's Point. His son was Sir William Phipps, who eventually became Royal governor of the new colony. Demeritt (1973, 1-3, 165) lists his name as Phips and his working dates in "Pemequid & Woolwich" from 1638-1654.

Pickering, F. B. & Co.   Lee   -1899-1900-
Tools Made: Farm Tools

Picket, Joseph**   Montville   -1850-
Tools Made: Blacksmith
Remarks: He was born in Raymond, Maine, and is listed in the 1850 census.

Pierce, C. S.**   ?   -1830-1840
Tools Made: Planes
Remarks: The Bob Jones collection has a 10 ½" beech double iron sash plane marked **C.S.PIERCE.** with no place name (Pollak 2001, 322). There is no relationship between this plane's mark and the later Cecil Pierce mentioned below.

Pierce, Cecil**   Southport   (d.1998)
Tools Made: Planes
Remarks: A late 20th century planemaker from Southport, who died in 1998. He was a frequent customer of the Liberty Tool Co. and purchased many of his plane blades there. The Davistown Museum is seeking a Cecil Pierce plane for its collections. He has written a book *Fifty Years a Planemaker and User* (Pierce 1994). The rosewood and applewood 12" smooth plane (photo) belongs to Rick Floyd.

Pierce, H. E.**   Belfast   -1854-
Tools Made: Wood Lathes
Remarks: Jeff Joslin indicates that, based on patent information, Milton Roberts and

H. E. Pierce were lathe-makers working as partners in 1854. By 1856, Roberts had a new partner, Isaac N. Feltch. All of their patents related to wood lathes for production use.

Pike, Oscar   Princeton   -1869-1871-
Tools Made: Rakes (Horse Rakes)

Pingree, L. F.**   Portland
Tools Made: Levels
Remarks: Two bench levels were listed in a Brown Auction Services catalog marked **L. F. PINGREE PORTLAND ME.** One level was 9" and had brass plates top and bottom and mahogany infill. The other was 8" long with all brass with three cutouts and a side porthole. Tim Daniels has located a 29 ½" long, 3 ¼" deep, and 1 3/8" wide level with full brass end plates about ¼" thick. He feels it looks to be contemporary with other makers from around 1830 – 1850.

Piscataquis Iron Foundry   Foxcroft   -1871-
Tools Made: First Plows

Pitman, Joseph A.**   Bangor   -1855-
Tools Made: Brass Founders and Finishers
Remarks: He is listed in the 1855 *Maine Registry and Business Directory.*

Place, William S.**   Charleston   1862-1879
Tools Made: Gunsmith and Clockmaker
Remarks: Demeritt (1973, 165).

Plimpton, Elias & Sons   Litchfield   -1855-1885-
Tools Made: Forks and Hoes
Remarks: They are listed in the 1855 *Maine Registry and Business Directory.* According to a "History of Litchfield" article "Elias Plimpton came to Litchfield from Walpole, Mass., in 1820, and established the hoe and fork manufactory which is known as E.Plimpton & Sons' Manufactory. The business is now carried on by his sons, A.W. and George Plimpton. Elias was born November 12, 1794; married Nancy Billings, July 16, 1820; died October 9, 1886. Nancy, his wife, born March 25, 1795; died October 15, 1885. Asa W., b. November 7, 1825. George, b. May 7, 1828" (*Kennebec Journal* 1897, 261).

Plummer, Silas   Lisbon Falls   -1850-1870-   (b.1822)
Tools Made: Wood Planes
Remarks: Silas first appears in the 1850 census as a carpenter and then again in the 1870 *Products of Industry* census as making $700 worth of plane stocks and tool chests.

These are rated as extremely rare by Pollak (2001, 325). He used the mark **S. PLUMMER | LISBON FALLS ME** with the name line curved.

Poland, John**   Montville   -1880-
Tools Made: Machinist Apprentice
Remarks: He is listed in the 1880 census.

Pomroy, Thomas   Bangor   -1851-
Tools Made: Plows

Pope, J. & Son   Manchester   -1869-1881-
Tools Made: Forks and Granite Wedges
Remarks: This company is listed in the 1881 *Maine Business Directory*.

Pope, Jacob**   Manchester   -1855-
Tools Made: Forks
Remarks: He is listed in the 1855 *Maine Registry and Business Directory*.

Porter, H.   Bangor   1835

Porter, J. W.   Sebago   -1849-
Tools Made: Farm Tools

Porter, Slyvanus**   Paris   -1855-
Tools Made: Pump and Block Makers
Remarks: He is listed in the 1855 *Maine Registry and Business Directory*.

Portland Wrench Co.   Portland
Tools Made: Wrenches
Remarks: Predecessor to the Diamond Wrench Co. George Short has provided this photograph of what he feels is one of their earlier wrenches. Also see H. A. Thompson.

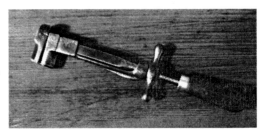

Potter, Elisha**   Bath   -1882-
Tools Made: Shipsmith
Remarks: He is listed in the 1882 *Maine Business Directory* on Water St.

Pratt, T. W.   Milo   -1862-
Tools Made: Handles (Shovel Handles)

Preble & Clark   Sullivan   -1873-

Tools Made: Adzes and Axes

Remarks: An ax made by this company marked **PREBLE & CLARK | L.GRAY** has
been reported to the museum (photograph courtesy
of Iron Horse Sculpture). L. Gray is probably an
owner's mark. More information is sought on this
downeast toolmaker.

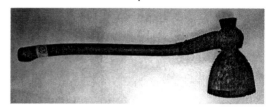

Preble, C. H. & Co.**   Sullivan   -1882-
Tools Made: Axes
Remarks: They are listed in the 1882 *Maine Business Directory*.

Proctor & Bowie Co.   Winslow
Tools Made: Axes
Remarks: The **PROCTOR & BOWIE CO. | WINSLOW, MAINE** mark appears on paper
labels applied to the axes.

Pulsifier, D. H.   Waldoboro   -1885-
Tools Made: Farm Tools

Quimby, Phineas**   Belfast   -1829-
Tools Made: Carriage maker
Remarks: Job White and Phineas Quimby received an 1829 patent for a "machine for
cutting veneers and panels for carriages." Information courtesy of Jeff Joslin.

Quinn, John**   Robbinston   -1856-
Tools Made: Shipsmith
Remarks: He is listed in the 1856 *Maine Business Directory*.

Quint, John   North New Portland   -1849-
Tools Made: Edge Tools

R., C. (See C. Record)

Ramsdell & Roth**   Bangor   1873-1874
Tools Made: Gunsmith
Remarks: Demeritt (1973, 165) notes this is John W. Ramsdell and Ernest Roth.

Ramsdell & Neal**   Bangor   1872-1878
Tools Made: Gunsmith
Remarks: Demeritt (1973, 67-72, 165) notes this is Charles V. Ramsdell and John H.
Neal.

Ramsdell, Charles S.*   Bangor   1887-1890-
Tools Made: Gunsmith
Remarks: Demeritt (1973, 165) notes his working address as Harlow & State Sts.

Ramsdell, Charles V.*   Bangor   -1850-1878   (d. 1886)
Tools Made: Gunsmith
Remarks: Demeritt (1973, 67-72, 165) notes his working address as East Market Square, Harlow St. from the 1850s to 1886 and then he worked later on State St. He was a partner in Ramsdell & Neal from 1872 – 1878 with John H. Neal. This partnership is listed with an address of 1 Harlow St.

Ramsdell, John W.*   Bangor   1860-1880-
Tools Made: Gunsmith
Remarks: Demeritt (1973, 165) notes his working address as East Market Square, Harlow & State Sts. He was partnered with Ernest Roth in Ramsdell & Roth from 1873-1874.

Randall, B.*   -1790-
Tools Made: Planes
Remarks: Pollak (2001, 334) states that this person is believed to be from a Freeport, ME, shipbuilding family. Examples marked **R. RANDAL** (with the "N" reversed") include two 10" beech molders with flat chamfers found in Maine.

Randel, Charles*   Montville   -1860-
Tools Made: Blacksmith
Remarks: He is listed in the 1860 census.

Ray & Osgood   Blue Hill
Tools Made: Coopers' Axes and Slicks
Remarks: DATM (Nelson 1999) lists this company as a maker of coopers' axes and states that it is not clear if Blue Hill was a city or brand name. Blue Hill, Maine, is about 50 miles from Bangor where Matthew Ray was working. A slick with the mark **RAY & OSGOOD | BLUE HILL | REFINED | CAST STEEL** has been reported to the Davistown Museum. Any information about the working dates and location of this company would be appreciated. See comments below on Matthew Ray.

Ray, Matthew   Bangor   -1855-1859-
Tools Made: Adzes and Edge Tools
Remarks: Matthew Ray marked his tools: **M.RAY | BANGOR** (Nelson, 1999). He is listed in the 1855 *Maine Business Directory* as an edge toolmaker. RootsWeb lists a

number of Ray and Osgood surnames in Blue Hill, Maine. These include: Ray, Matthew b: 24 Nov 1784 in Surry, ME (Surry is the next town over from Blue Hill). Several of his children from his second wife (1833f.), were born in Bangor. This, probably is the same Ray as in Ray & Osgood. It is unknown who Osgood was.

Raymond, F. & Co.**　　Lewiston　　1882
Tools Made: Gunsmith
Remarks: Demeritt (1973, 165).

Raymond, W.**
Tools Made: Planes
Remarks: Pollak (2001, 336) lists an example, marked **W. RAYMOND**, of a 9 7/16" beech, 3/16" center bead, with no boxing, and **7** imprinted on the heel, found in Maine.

Read, George T.**　　Belfast　　1876-1900
Tools Made: Gunsmith
Remarks: Demeritt (1973, 165).

Record, C.**　　-1790-1820-
Tools Made: Planes
Remarks: Pollak (2001, 337) lists the initial group, marked **C.R.** with a leaf motif, as a 9" birch round, a 9 5/8" beech quarter round, and a 8 3/4" birch small cove. The quarter round and complex molders have ogee molded shoulders. All have small flat chamfers and were found in Maine. The full name examples, marked **C.RECORD** with a leaf motif, are a large hollow and an astragal, both 9 3/8" beech with round chamfers, found together in Maine.

Record, S. J., Co.　　Norway
Tools Made: Axes
Remarks: His mark: **S.J.RECORD CO. | NORWAY, MAINE** was printed on a paper label, but not on the ax.

Redlon & Simms　　Portland　　-1869-
Tools Made: Brushes (Brooms)

Reed, Jason**　　Bucksport　　-1855
Tools Made: Pump and Block Makers
Remarks: He is listed in the 1855 *Maine Registry and Business Directory.*

Reed, W.**　　-1800-
Tools Made: Planes

Remarks: Pollak (2001, 338) lists the examples, marked **W:REED**, of a 1/2" bead and two birch complex molders, all 9 5/8" with flat chamfers found in Maine.

Reynolds, Charles H. & Co.**   Lewiston   -1855-
Tools Made: Machinist Tools and Saw Blades for Rotary Sawmills
Remarks: They have an advertisement on page 21 of the 1855 *Maine Business Directory* Advertising Supplement (figure).

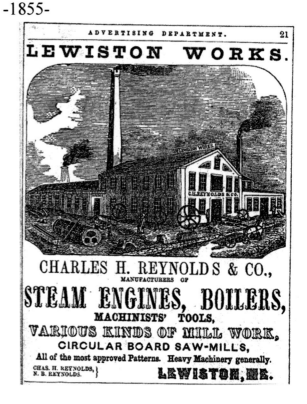

Reynolds, Oliver M.**   Lubec   -1856-
Tools Made: Shipsmith
Remarks: He is listed in the 1856 *Maine Business Directory.*

Rice & Miller   Bangor
Tools Made: Saws
Remarks: Their mark **RICE & MILLER | MECHANICS FAVORITE | BANGOR. ME** had many variations of letter style, size and positioning.

Rice, D.**   -1810-1830-
Tools Made: Planes
Remarks: Pollak (2001, 340) lists examples, marked **D.RICE**, of a 9 1/2" beech square rabbet with shallow flat chamfers and a 13 1/2" beech jack-style square rabbet with a centered open tote and round chamfers, both found on the coast of Maine.

Rice, Jesse, Jr.**   Prospect   -1855-
Tools Made: Pump and Block Makers
Remarks: He is listed in the 1855 *Maine Registry and Business Directory.*

Rice, M. E.   Stetson
Tools Made: Unknown

Rich, William**   Saco   1741-1743
Tools Made: Gunsmith
Remarks: Demeritt (1973, 165).

Rich, Zebediah T.**   Tremont (Bass Harbor)   -1856-
Tools Made: Shipsmith
Remarks: He is listed in the 1856 *Maine Business Directory.*

Richardson**   Brunswick   -1795-
Tools Made: Nails
Remarks: Richardson had a shop in Brunswick where he made nails for clapboards and shingles from iron hoops taken from rum barrels, and, as rum barrels were plentiful, he had no difficulty obtaining hoops sufficient for his purpose (Wheeler 1878, 582).

Richardson, G. E.   Hiram   -1849-
Tools Made: Edge Tools

Richardson, Jason E.**   Skowhegan   -1870-   (d. 1889)
Tools Made: Gunsmith
Remarks: Demeritt (1973, 165).

Richardson, William   Glenburn   -1869-
Tools Made: Plows

Richmond, John W.   ? -1856-   (b. 1828 d. 1866)
Tools Made: Planes
Remarks: Pollak (2001, 343) states that the mark **J.W.RICHMOND** is thought to be Capt. John W. Richmond (b. 1799, in Bridgewater, MA, d. 1867) or his son, John W. Richmond, both of whom were listed in the 1856 Maine directory as J. W. Richmond & Co. house builders. The Bob Jones collection has two planes with this mark, both found in Maine, a 13 ½" match grooving plane and a cornice ogee plane.

Richon, Aaron**   Verona   -1867-
Tools Made: Splitting Knives
Remarks: He is listed in the 1867 *Maine Business Directory.*

Ricker, G. B.**   Cherryfield   -1835-1880-
Tools Made: Chisels, Gouges, and Drawknives
Remarks: The Museum has numerous G. B. Ricker tools in the MIII and MIV collections, including the drawknife (ID# 62504T1) in the photo and (ID#s 40501T2, 12801T10, and 12801T9). The Amaziah Ricker House, built in Federal style in 1803, is located in Cherryfield, Maine. It is the second oldest house in the historic district. The Rickers were the first blacksmiths in the area, and the shop was in the adjacent building. The Masonic Lodge organized and met in this house. The 1856 *Maine Business Directory* lists Benj. G. Ricker as a shipsmith in Cherryfield. The 1860 Cherryfield census lists

Benjamin G Ricker, blacksmith, age 50. He has 5 daughters listed and two sons: William Ricker, Blacksmith, age 23 and Amaziah Ricker, age 17, no occupation. In 1870, all three are still listed, with Amasiah, age 26, working in a blacksmith shop. (We believe that Amaziah was spelled incorrectly as Amasiah.) In 1880, all three are listed as blacksmiths. The Ricker clan was a major source of edge tools for local shipbuilders. It is unknown how many generations of Rickers preceded Benjamin and when and where they began careers as artisans in hot iron.

Ricker, T. H. & Sons   Harrison   -1862-1885-
Tools Made: Carpenter Tools, Farm Tools, and Plows
Remarks: Among the tools they made were various machines for sawing, planing, shingle making, stave making, etc.

Rideout, E. N.**   Rockland   -1867-
Tools Made: Edge Tools
Remarks: He is listed in the 1867 *Maine Business Directory*.

Rider, Perry B.   Bangor   -1834-1848-   (b. 1808)
Tools Made: Wood Planes
Remarks: After 1848, Rider moved to Boston. If he continued to make planes there, it is not known. He used the mark **P.B.RIDER | BANGOR**. The Bob Jones collection includes the following planes with this mark: a 9 ½" beech boxed side beading plane with round chamfers, a 9 3/8" birch rabbet skewed plane also stamped **E.MANCHESTER** and the only one not stamped Bangor, a 9 ½" birch fillister plane, a 9 7/8" birch fillister plane, a 9 ½" beech coping plane, a double boxed mitered sash plane also stamped **ME 1840**, and a 7" lignum vitae smooth plane. Pollak (2001, 173) lists a 7" lignum vitae smoothing plane marked **J. W. GOODHUE | BANGOR. ME.** made by P. B. Rider/Bangor.

Rider, T. J.   Thomaston   -1861-
Tools Made: Wood Planes
Remarks: The DATM (Nelson 1999) listing does not report a town, state, or working dates for this planemaker, but suggests that the mark **T.J.RIDER.** may belong to a Thomas J. Rider. According to Pollak (2001), the imprint has been found on a beech molding plane. "The plane's iron is incised **BANGOR ME**. Probably Thomas J. Rider, born in 1806, who was a joiner in Thomaston, ME, and also a guard at the state prison in Thomaston. Thomaston is some 65 miles south of Bangor. The imprint has also been reported struck four times in the form of a square on a 22" long beech fore plane" (Pollak

2001, 344). The Bob Jones collection has a double-boxed center bead also marked **M.COPELAND & CO | WARRANTED**. Melvin Copeland worked in Huntington, MA, from 1842 until 1855 (Pollak 2001, 103-4). Rider might have been a hardware merchant who marked planes made by others.

Ridgeway & Thornhill   Portland   -1869-1871-
Tools Made: Files
Remarks: Ridgeway has also been recorded as Ridgway, without the "e".

Ridgeway, Samuel**   Gorham   -1882-
Tools Made: Edge Tools
Remarks: He is listed in the 1882 *Maine Business Directory*.

Rig, D.**   -1800-
Tools Made: Planes
Remarks: Pollak (2001, 344) lists a 9 1/2" beech molder with flat chamfers, marked **D.RIG**, found in Maine.

Ripley, G. W.**   Paris   -1867-
Tools Made: Edge Tools
Remarks: He is listed in the 1867 *Maine Business Directory*.

Ritchie, John (see Frederick W. Ritchie)

Ritchie, Frederick W.   Vanceborough   1884
Tools Made: Bevels and Squares
Remarks: The mark **AUG.19,1884** found on a combination bevel and try square corresponds to a patent that Ritchie held; whether he was the actual maker of the tool is not known. Tom Lamond notes that he held Pat. No. 303,666, for a combination tool (combination square/bevel), Aug. 19. 1884. Two thirds was assigned to John Ritchie and Michael L. Ross of the same place.

Robbins, Leon**   Bath   -1985-2001   (d. 2008)
Tools Made: Planes
Remarks: Leon was an active toolmaker making small hand planes, such as violin makers' and patternmakers' planes, as recently as 2001. His tool shop in Bath (Crown Plane Co.) played a role in inspiring the organization of the Jonesport Wood Co. in West Jonesport (1970). He sold the Crown Plane Co. to James White in 2001. The Davistown Museum has numerous examples of his oeuvre, some recently donated by his daughter (ID#s 52408t1, 52408t2, 52408t3, 52408t4, 52408t5, 52408t6, and 52408t7), in the IR

collection. Pictures of them are available on the Davistown Museum website. The Bob Jones collection has one of his tiger maple veneer single boxed side bead planes.

Roberts, Milton**   Belfast   -1854-1856-
Tools Made: Wood Lathes
Remarks: Jeff Joslin indicates that, based on patent information, Milton Roberts and H. E. Pierce were lathe makers working as partners in 1854. By 1856, Roberts had a new partner, Isaac N. Feltch. All of their patents related to wood lathes for production use.

Roberts, Nicholas**   Yarmouth   -1881-1882-
Tools Made: Edge Tools
Remarks: He is listed in the 1881 and 1882 *Maine Business Directories*.

Robinson**   Calais   -1855-
Tools Made: Wheelwright
Remarks: He is listed in the 1855 *Maine Registry and Business Directory*.

Robinson Co.   Presque Isle   -1899-1900-
Tools Made: Farm Tools

Robinson, E.**   Skowhegan   1898
Tools Made: Gunsmith
Remarks: Demeritt (1973, 165).

Robinson, Greenlief**   Bridgton   1885
Tools Made: Gunsmith
Remarks: Demeritt (1973, 165).

Robinson, James E.   Portland   -1855-
Tools Made: Farm Tools

Robinson, Joseph   Warren   -1867-1871-
Tools Made: Axes and Shipsmiths' Tools
Remarks: He is listed in the 1867 and 1869 *Maine Business Directories*.

Robinson, Lewis C.**   Durham   -1882-
Tools Made: Edge Tools
Remarks: He is listed in the 1882 *Maine Business Directory*.

Rolf, James**   Scarborough   -1832-
Tools Made: Axes
Remarks: He is listed in *American Axes* by Henry J. Kauffman (1972).

Rollin, B. R.**   Weld   -1855-
Tools Made: Edge Tools
Remarks: He is listed in the 1855 *Maine Business Directory*.

Rollins, Joseph L.**   Augusta
Tools Made: Screwdrivers
Remarks: He was issued patent 902608, 11/3/1908, information courtesy of C.D. Fales.

Rollins, T. H.**   Detroit   -1870-
Tools Made: Saws
Remarks: He had patent 99596 on 2/8/1870 for a saw frame. Source: Graham Stubbs.

Ropes, David   Saccanapa   1832-
Tools Made: Cutlery
Remarks: There does not appear to be any town of this name. When questioned about it, the Yarmouth Historical Society responded that "Saccarappa is the most used spelling for this area/village in the town of Westbrook. It's not used today, but it was just upstream on the Presumpscot River from Cumberland Mills (another village in Westbrook and site of S.D. Warren's big mill). You will find it on old county atlases, and in the Maine Register (within Westbrook)."

Ross, Andrew B.   -1869-
Tools Made: Plows

Ross, Benjamin D.**   Portland   1823
Tools Made: Gunsmith
Remarks: Demeritt (1973, 165).

Ross, Michael L. (See Frederich W. Ritchie)

Roth, Ernest**   Bangor   1874
Tools Made: Gunsmith
Remarks: Demeritt (1973, 165) notes he worked in partnership with John W. Ramsdell in Ramsdell & Roth.

Rounds, Mark**   Scarboro, Newcastle, Saco, Falmouth (now Portland)   1681-1720
Tools Made: Gunsmith
Remarks: Demeritt (1973, 165) notes his dates as Scarboro (1681), Newcastle (1784), Saco (1699), and Falmouth (1715-1720).

Rounds, Nathaniel**   Waterford, South Waterford   -1855-1867-   (b. 1799 d. 1868)
Tools Made: Axes and Edge Tools
Remarks: He was born in Buxton in 1799 and moved to Waterford Lower Village in 1816, where he died in 1868. He was a blacksmith and used the mark **N. Rounds**, information courtesy of Trevor Robinson of Amherst, MA. He is listed in the 1855 and 1856 *Maine Business Directories* as an edge toolmaker in Waterford and in the 1867 *Maine Business Directory* as making axes in South Waterford and Waterford.

Rouse, William**   Bath   -1854-
Tools Made: Blocks
Remarks: Baker (1973, 436) lists William Rouse as one of three block-makers in Bath.

Rowe, O. P.**   Pittston   -1879-1882-
Tools Made: Axes
Remarks: He is listed in the 1879, 1881, and 1882 *Maine Business Directories*.

Rowe, Webster**   Skowhegan, Hallowell   -1850-1900   (d.1918)
Tools Made: Gunsmith
Remarks: Demeritt (1973, 165) notes him in Skowhegan (1850s- 1860s), Hallowell (1870s – 1880s) and Skowhegan (1891-1900).

Russell, Dexter W.**   Camden   -1881-1882-
Tools Made: Shipsmith
Remarks: He is listed in the 1881 and 1882 *Maine Business Directories*.

Russell, John   Cambridge   -1849-
Tools Made: Edge Tools

Russell, L.**   Phillips   -1855-
Tools Made: Edge Tools
Remarks: He is listed in the 1855 *Maine Business Directory*.

Russell, T. H.**   Wellington   -1867-
Tools Made: Axes and Chisels
Remarks: He is listed in the 1867 *Maine Business Directory* as an ax-maker. Rick Floyd has given us the two photographs of a socket chisel marked **T.H.RUSSEL**.

Saco River   -1835-
Tools Made: Rules

Remarks: According to the DATM (Nelson 1999), "An incidental comment on a separate data entry mentions a 'Saco River rule dated 1835'. The rule type is a board/log measure and the comment notes that, while all octagonal rules are not Saco River ones, all Saco River rules are octagonal. It is not clear whether Saco River is a maker, a city, an area, a style, etc."

Safford & Co.**   Augusta
Tools Made: Gunsmith
Remarks: Demeritt (1973, 165). See Charles Henry Safford.

Safford & Parsons**   Augusta   1874-1875
Tools Made: Gunsmith
Remarks: Demeritt (1973, 165) notes this is Charles Henry Safford and John H. Parsons.

Safford, Charles Henry**   Augusta   1871-1880
Tools Made: Gunsmith
Remarks: Demeritt (1973, 165) notes him as working with John H. Parsons as Safford & Parsons (1874-1875) and also as Safford & Co.

Sampson, Abel   Portland   1820-1846   (b. 24 Aug. 1790, d. 1883)
Tools Made: Tools and Wood Planes
Remarks: Born in Turner, Maine, Sampson was a joiner, toolmaker, and planemaker in Portland from 1823-37, according to Pollak (2001, 354). The DATM (Nelson 1999) also indicates that, in 1846, he moved to Lawrence, MA, where he was a machinist. By 1869 he was back in Maine as a carpenter and joiner. He may have also worked in New York City early on in his life around 1817-18. He used the mark **A. SAMPSON | PORTLAND** with a curved city line, and, at other times, he left the city name out. Pollak rates both his marks as "rare." For further information see "ABEL SAMPSON: Maine's Privateer turned Planemaker" by Dale Butterworth & Bennet Blumenberg in *The Chronicle* (Butterworth 1992). The Bob Jones collection has a 9 ½" beech dado plane and a 9 ¼" beech Grecian ovolo plane with his mark. John Harper has a skew rabbet from the 1823-37 time period.

Sanborn, C. W. & Co.   Morrill
Tools Made: Rakes (Horse Rakes)

Sanders, D. T. & Sons**   Greenville
Tools Made: Axes
Remarks: They are listed in Appendix 2 of Klenman's *Axe Makers of North America* without further discussion (Klenman 1990, 101).

Sandfords, G. W.**   Manchester   -1879-
Tools Made: Knives
Remarks: He is listed in the 1879 *Maine Business Directory*.

Sargent, C. A.**   Lincoln   1874-1900
Tools Made: Gunsmith and Blacksmith
Remarks: Demeritt (1973, 165).

Saunders, Austin**   Bucksport   -1869-
Tools Made: Shipsmith
Remarks: He is listed in the 1869 *Maine Business Directory*.

Saunders, H. P.   Kingfield   -1879-
Tools Made: Rakes

Savage, Chandler**   Concord   1880-1881
Tools Made: Smith
Remarks: He is listed by Hoyt (1881).

Savage, Stillman**   Temple   1877
Tools Made: Gunsmith
Remarks: Demeritt (1973, 165).

Sawyer, Joshua W.**   Portland   1846-1850
Tools Made: Gunsmith
Remarks: Demeritt (1973, 165).

Sawyer, R. H.   Gardiner   -1869-1871-
Tools Made: Saws

Scammon, Thomas M.**   Yarmouth   -1869-
Tools Made: Shipsmith
Remarks: He is listed in the 1869 *Maine Business Directory*.

Scammons, Joseph**   Franklin   -1855-
Tools Made: Wheelwright
Remarks: He is listed in the 1855 *Maine Registry and Business Directory*.

Schwartz, Michael   Bangor   1843-1880-

Tools Made: Files, Rules, Saws, Squares, and Other Tools

Remarks: After working in the Boston area from 1839-43, where he may or may not have been involved with making tools, Schwartz moved to Bangor, Maine, where he became quite involved with the tool business, working with at least five different companies, some concurrently. He began making saws and continued that business for quite some time, but he also made files briefly, worked with Job Collett around 1855, and made rules and squares as a partner in Darling & Schwartz from before 1852 to 1868. His other tool activities included making shingle machines and patenting a hoop splitter/shaver, which may not have been made. There is also a possibility that he made the Evans iron circular plane (28 Jan. 1862 and 22 March 1864 patents). M. Schwartz's Sons succeeded him in the saw business by 1888 (Nelson, 1999).

Schwartz's Sons, Michael   Bangor   -1888-1914

Tools Made: Saws

Remarks: This company succeeded Michael Schwartz.

Scott, H.   Plymouth   - 1869-1881-

Tools Made: Axes

Remarks: He is listed in the 1869 and 1881 *Maine Business Directories*. Rick Floyd has a mast ax signed **Scott**.

Searles, David C.**   Livermore Falls   1897

Tools Made: Pruning Tools

Remarks: He has Pat. No. 579,083 for a reversible pruning tool, Mar. 6, 1897 (Tom Lamond, personal communications).

Searsport Manufacturing Co.**   Searsport   -1855-
Tools Made: Pump and Block Makers
Remarks: They are listed in the 1855 *Maine Registry and Business Directory*.

Seavey & Chalmer**   East Machias   -1856-
Tools Made: Edge Tools
Remarks: This company is listed in the 1856 *Maine Business Directory* as an edge toolmaker.

Seavey, Charles H.*   Machias, East Machias   -1856-
Tools Made: Axes
Remarks: He is listed in the 1856 *Maine Business Directory* as an edge toolmaker located in East Machias, while Yeaton (2000) lists C. H. Seavey as an ax-maker in Machias with no dates.

Segar, Joseph**   Prospect   -1855-
Tools Made: Edge Tools
Remarks: He is listed in the 1855 *Maine Business Directory*.

Senter, William & Co.   Portland   -1873-1888   (b. 11 Oct. 1813, d. 22 Dec. 1888)
Tools Made: Scientific Instruments
Remarks: From 1836-70, Senter was part of Lowell & Senter. In 1873, he worked under his own name, but after that he is listed with "& Co."

Sevey & Chaloner**   East Machias   -1856-
Tools Made: Shipsmith
Remarks: They are listed in the 1856 *Maine Business Directory*.

Sevey, Charles H.**   East Machias   -1856-
Tools Made: Shipsmith
Remarks: He is listed in the 1856 *Maine Business Directory*.

Sewall, S.**   York   -1745-1760-   (b. 1724, d. 1814)
Tools Made: Planes
Remarks: Pollak (2001, 367) states that this is possibly Samuel Sewall of York, ME who was a joiner, bridge-builder, and carpenter, active from 1745-1760. An example is a wedge-locked slide-arm filletster of cherry with arms set into the fence with iron rivets.

Seward, W.**   Bangor
Tools Made: Planes

Remarks: The Bob Jones collection contains a 9 ½" skewed rabbet plane marked
I.COOMBS | BANGOR | W.SEWARD.

Seymour Mfg. Co.* Oakland 1923-1924
Tools Made: Axes

Shaw & Maynes* Limestone 1883-1891
Tools Made: Axes

Shaw & Tenney, Inc.** Orono 1858-to date
Tools Made: Boat Hooks, Rowing Hardware, Oars, and Paddles
Remarks: Still in existence, this company uses methods and patterns which have been essentially unchanged throughout its history. See www.shawandtenney.com.

Shaw, Edgar** Portland 1890
Tools Made: Unknown
Remarks: He has Pat. No. 428,421 for a fastening device for tools (hammer wedge), May 20, 1890 (Tom Lamond, personal communications).

Sherman & Co.* Belfast 1886-1896
Tools Made: Axes
Remarks: See Sherman & Thompson.

Sherman & Thompson Belfast -1882-1885
Tools Made: Axes and Adzes
Remarks: Their mark was: **S & T | BELFAST.ME** with the initials in an oval outline. This company is listed as being on Goose River in the 1882 *Maine Business Directory.* Sherman & Co. followed this partnership and was in operation until 1896.

Sherman, Bridges Curtis** Liberty -1850-1880-
Tools Made: Blacksmith
Remarks: He is listed in the 1850 - 1880 censuses. He was living in the household of Abiel Sherman in 1850 and was the head of his own household by 1860.

Sherman, G. A. & Co.** Bucksport -1869-
Tools Made: Shipsmith
Remarks: They are listed in the 1869 *Maine Business Directory.*

Sherman, John H. & Co.** Bucksport -1856-
Tools Made: Shipsmith
Remarks: They are listed in the 1856 *Maine Business Directory.*

Sherman, Walter H.* Liberty -1880-
Tools Made: Blacksmith
Remarks: He is the son of Bridges Curtis Sherman. He is listed in the 1880 census.

Sides, Andrew* Waldoboro 1856-1869-
Tools Made: Shipsmith
Remarks: He is listed in the 1856 *Maine Business Directory*. The 1869 *Business Directory* lists Andrew Sides & Son.

Sidelinger, G. B.* Rockport, Camden -1881-1882-
Tools Made: Shipsmith
Remarks: He is listed in the 1881 and 1882 *Maine Business Directories*.

Silsbee, John* Bucksport -1855-
Tools Made: Wheelwright
Remarks: He is listed in the 1855 *Maine Registry and Business Directory*.

Simpson Scythe & Axe Co.* Oakland -1880-
Tools Made: Axes
Remarks: Lamond (2007b).

Simpson, Isiah H.* Brunswick 1895
Tools Made: Gunsmith
Remarks: Demeritt (1973, 165) notes he held patent 542,540.

Simpson, James S.* Ellsworth -1855-
Tools Made: Pump and Block Makers
Remarks: He is listed in the 1855 *Maine Registry and Business Directory*.

Simpson, Laforest* Waterville 1861
Tools Made: Gunsmith
Remarks: Demeritt (1973, 165).

Simpson, William* Benton -1856-
Tools Made: Edge Tools
Remarks: He is listed in the 1856 *Maine Business Directory*.

Sinclair, Albert Waterville -1869-
Tools Made: Brushes (Brooms)

Sinclair, Benjamin**   Pembroke   -1856-
Tools Made: Shipsmith
Remarks: He is listed in the 1856 *Maine Business Directory*.

Sinclair, R. W.   Houlton
Tools Made: Rules (Log Rules)

Skowhegan Broom Co.   Skowhegan   -1869-
Tools Made: Brushes (Brooms)

Slocomb, Caleb*   New Fairfield   1867-1888
Tools Made: Axes
Remarks: It is unknown if there was a connection with C. E. Slocumb or if this was the same man and one spelling of the name is incorrect.

Slocomb, J. L.*   New Portland
Tools Made: Axes

Slocumb, C. E.   Fort Fairfield   -1867-1871-
Tools Made: Axes
Remarks: His name was also reported as "Slocum." Yeaton (2000) spells his name as "Slocomb," leading to the suggestion that this Slocumb may have some connection with the other Slocombs. He is listed as Slocumb in the 1867 and 1869 *Maine Business Directories*.

Small, Oliver*   Otisfield   -1874-1882-
Tools Made: Axes
Remarks: He is listed in the 1882 *Maine Business Directory*.

Smallwood**   ?   -1835-
Tools made: Framing Squares
Remarks: The Davistown Museum has a Smallwood framing square in the MIII collection (ID# 63001T2).

Smart, Alfred**   Pittston   -1856-
Tools Made: Edge Tools
Remarks: He is listed in the 1856 *Maine*

Business Directory. The Davistown Museum has a framing chisel (ID# 51606T1) marked **A. SMART** in the MIV collection (photo).

Smart, Charles H.   Brighton   -1867-1871-
Tools Made: Axes
Remarks: He is listed in the 1867 and 1869 *Maine Business Directories*.

Smart, William H.   Swanville   -1862-
Tools Made: Ax Handles

Smith, A. T.**   Gardiner
Tools Made: Sleigh
Remarks: The remains of an old sleigh were found by Michael Jarrett. On one of the pieces of wood was a brass plate stating **A. T. SMITH GARDINER MAINE**. It is possible that he made carriages too.

Smith, Charles C.   Dover   -1860-1881-
Tools Made: Edge Tools, Blacksmith, and Gunsmith
Remarks: He is listed in the 1879 and 1881 *Maine Business Directories* for making edge tools. Demeritt (1973, 165) lists him as a gunsmith and blacksmith.

Smith, D. W.**   Machias   -1918-
Tools Made: Buck Saws
Remarks: He had patent 1253898 on 1/15/1918 for a buck saw. Source: Graham Stubbs.

Smith, Elliot**   Bath   -1850-
Tools Made: Shipsmith
Remarks: Information from Baker (1973) *Maritime History of Bath, Maine*.

Smith, Jesse**   Bingham   -1850-1870-
Tools Made: Blacksmith
Remarks: He is listed by Moore (1840). The Old Canada Road Historical Society web page has photographs of the Jesse Smith Sr. and Jr. homestead. They were blacksmiths since 1850
(http://www.rootsweb.ancestry.com/~meocrhs/smith_collection/smith_one.htm).

Smith, Jesse & Son   Bingham   -1855-1856-
Tools Made: Edge Tools
Remarks: They are listed in the 1855 and 1856 *Maine Business Directories* as edge toolmakers.

Smith, Joel   Greenville   -1874-1883-
Tools Made: Axes
Remarks: He is listed in the 1879 and 1881 *Maine Business Directories*.

Smith, Q. D.**   Bath   1856
Tools Made: Gunsmith
Remarks: Demeritt (1973, 165).

Smith, R. H.**   Bingham, Moscow   1877
Tools Made: Gunsmith
Remarks: Demeritt (1973, 165).

Smith, Reuben**   Montville   -1850-
Tools Made: Blacksmith
Remarks: He is listed in the 1850 census.

Smith, Rueben W.**   Bucksport   1885-1900   (d. 1905)
Tools Made: Gunsmith
Remarks: Demeritt (1973, 165).

Smith, William G. & Son**   Bucksport   -1850-1884   (d. 1884)
Tools Made: Gunsmith
Remarks: Demeritt (1973, 165).

Smith, William H. & Son**   Bath   1867-1881
Tools Made: Gunsmith and Sporting Goods
Remarks: Demeritt (1973, 165).

Snow & Nealley*   Bangor   1864-to date
Tools Made: Axes and Timber Harvesting Tools
Remarks: This is one of Maine's most productive toolmakers. The Davistown Museum has a Snow & Nealley pulp hook (ID# 31908T39), pickaroon (ID# 4106T11), and peavey (ID# 61204T1) in the IR collection. The peavey is also marked **E. MANSFIELD & Co.** A photo of it is under the "Mansfield" listing in this registry. Visit their website at www.snowandnealley.com.

Snow, Francis A.**   Sedgwick   -1856-
Tools Made: Shipsmith
Remarks: He is listed in the 1856 *Maine Business Directory*.

Snow, Kenny**   Bucksport   -1855-
Tools Made: Pump and Block Makers
Remarks: He is listed in the 1855 *Maine Registry and Business Directory*.

Snowdeal, C. T.   Thomaston   -1850-1877
Tools Made: Wood Planes
Remarks: The imprint **C.T.SNOWDEAL** (arranged in a semi-circle), sometimes with **ME | THOMASTON** below it, has been reported by Pollak (2001, 384) on a 9 3/8" long beech spar plane. Pollak (2001) also reports a dado plane in the Bob Jones collection that is additionally stamped **GLADWIN & APPLETON | BOSTON**. Porter A. Gladwin and Thomas L. Appleton made planes in Chelsea, MA, from 1873-1877 (Pollak 2001, 169). Bob Jones also has a ships' hollow plane with the Snowdeal Thomaston mark.

Soule, Lewis S.   Waldoboro   -1849-1854-   (b.1813)

Tools Made: Wood Planes
Remarks: Soule was a joiner who also made doors, sashes, and blinds. He left either the imprint **L.S.SOULE | WALDOBORO | ME.** or simply **LS** with an abstract evergreen tree between the letters on his planes. The Davistown Museum has a Soule curved beading plane (ID# 32708T58) and a smooth plane (ID# TBW1005) in the MIV collection (photo). The Bob Jones collection includes a number of planes with this imprint that also have other factory imprints (**UNION FACTORY, MULTIFORM | MOULDING PLANE Co., D. COPELAND | HARTFORD**). It is possible that he was also a hardware merchant.

Soule, W. H.**   Portland   1896
Tools Made: Gunsmith
Remarks: Demeritt (1973, 165).

Spafford, Amos*   Buxton   -1874-
Tools Made: Axes

Spalding, Abel**   North Buckfield   -1840-1880-
Tools Made: Gunsmith
Remarks: Demeritt (1973, 53-4, 165) notes he moved to Swanton, Ohio, around 1874.

Sparrow, J.   Portland   ca. 1827-
Tools Made: Wood Planes
Remarks: J. Sparrow of Portland held a 26 Dec. 1827 patent for a "sliding plane turner." A compass plane has also been reported marked **J. Sparrow** without a city.

Sparrow, William**   Portland   -1855-
Tools Made: Agricultural Implements
Remarks: He is listed in the 1855 *Maine Registry and Business Directory*. It is possible that he is related to the J. Sparrow planemaker c. 1827 also of Portland.

Spear & Billings (see Billings & Spear)

Spear, C. A.**   Warren
Tools Made: Planes
Remarks: The Davistown Museum has a

razee plane (ID# 080704T1) marked **C A Spear** in the MIV collection (photo).

Spear, D. Howard**   Bath   -1886-   b.1849
Tools Made: Shipsmith and Blacksmith
Remarks: In 1886, he joined with Elijah F. Sawyer and Captain John R. Kelley in the shipbuilding firm Kelley, Spear & Company. He learned the blacksmith trade from his father, George J. Spear. He was a "master blacksmith on a number of noted Bath-built square-riggers including the *John R. Kelley* and the *Benjamin F. Packard*" (Baker 1973, 626).

Spear, George J.**   Bath   -1850-1881
Tools Made: Blacksmith and Shipsmith
Remarks: He was a smith for E. & A. Sewall (Baker 1973, 626). He is listed in the 1881 *Maine Business Directory* as a shipsmith.

Spencer, Samuel**   Industry-West Mills   1880
Tools Made: Gunsmith
Remarks: Demeritt (1973, 165).

Spiller Axe & Tool Co.**   Oakland   1826-1965
Tools Made: Axes
Remarks: This company is listed in Kauffman (1972) as working in 1928. Lamond (2007b) gives the dates 1826-1965.

Spiller, Mark D. & Sons*   Oakland   1926-1965
Tools Made: Axes

Spofford, Josiah**   Portland   1846-1860
Tools Made: Gunsmith
Remarks: Demeritt (1973, 165).

Sprague, John 2d**   Charlotte   -1855-
Tools Made: Wheelwright
Remarks: He is listed in the 1855 *Maine Registry and Business Directory*. It is unclear whether "2d" means he is John Sprague II.

Springate, E. A.**   Calais   -1855-
Tools Made: Pump and Block Makers
Remarks: He is listed in the 1855 *Maine Registry and Business Directory*.

Springer, R.*   Portland   1903-1910
Tools Made: Axes

Sprocket Wrench Co.**   Searsport   -1900-
Tools Made: Wrenches
Remarks: Rick Floyd has a wrench marked **PAT JAN 13 1900 | SPROCKET WRENCH CO SEARSPORT ME** (photos). According to the Directory of American Tool and Machinery Patents (www.datamp.org), this is not a valid patent date, and the correct date is Feb. 13, 1900 for patent #643,520 issued to Charles H. Monroe of Searsport, ME. A 1923 Shapleigh Hardware Co. catalog listing for the wrench was reprinted in the *MVWC Newsletter* (Missouri Valley Wrench Club 1987, 6). Another example (#128) is shown in Schulz's (1989) *Antique & Unusual Wrenches*. Both are marked **SPROCKET WRENCH || S & Co. N.Y.** The chain was long enough to wrap around most of a bicycle chain wheel, and that is what it was designed for.

Sprowl, Alfred P.**   Narraguagus - Cherryfield   -1860-1878
Tools Made: Gunsmith
Remarks: Demeritt (1973, 165).

Standard Framer Co.   Augusta
Tools Made: Other
Remarks: The tool they made was a "framer;" its exact identity and use is unknown.

Standish, Pierce & Co.   Ellsworth   -1869-
Tools Made: Farm Tools

Stanley, Frank**   Dixfield   -1870-
Tools Made: Gunsmith
Remarks: Demeritt (1973, 165).

Stanwood & Noyes**   Portland   -1869-
Tools Made: Shipsmith
Remarks: They are listed in the 1869 *Maine Business Directory* at 173 Commercial St.

Stanwood, Charles**   Portland   -1869-1882-
Tools Made: Shipsmith
Remarks: He is listed in the 1869 and 1881 *Maine Business Directory* at 171 Commercial St. and at 262 Commercial St. in 1882.

Stanwood, G. M. & Co**   Portland   -1881-
Tools Made: Shipsmith
Remarks: They are listed in the 1881 *Maine Business Directory* at 171 and 173 Commercial St.

Stanwood, George M.**   Portland   -1882-
Tools Made: Shipsmith
Remarks: He is listed in the 1882 *Maine Business Directory* at 261, 263, and 265 Commercial St.

Stanwood, Colonel William**   Brunswick   End of Rev. War to 1790   (b.1752 d.June 1829)
Tools Made: Blacksmith and Shipbuilder
Remarks: He took on James McFarland as a blacksmith apprentice. McFarland took over the business in 1790 (Wheeler 1878, 578, 808, 854).

Staples & Philbrook**   Bangor   1875
Tools Made: Gunsmith
Remarks: Demeritt (1973, 165) notes this was a partnership of Charles G. Staples and Francis J. Philbrook.

Staples, A.**   Portland
Tools Made: Axes
Remarks: He is listed in Appendix 2 of Klenman (1990, 102) without further discussion.

Staples, Charles G.**   Bangor   1875
Tools Made: Gunsmith and Machinist
Remarks: Demeritt (1973, 165) notes he was partnered with Francis J. Philbrook in Staples & Philbrook.

Staples, James Forest   Portland   -1849-1856-
Tools Made: Drawknives and Edge Tools
Remarks: A drawknife, assumed to be made by James Forest Staples, has been reported marked **J. Staples, Portland**. He is listed in the 1855 and 1856 *Maine Business Directories*.

Starr, R. C.**   Thomaston   (b. 1779 d. 1862)
Tools Made: Planes
Remarks: This imprint probably belonged to Reverend Robert C. Starr. He was born in 1779 in Massachusetts, lived in Thomaston, ME, and was a joiner before becoming a Baptist minister. His imprint is embossed and reads **R.C.STARR.** (Pollak 2001, 392). The Bob Jones collection includes a 10" coping plane.

Stenchfield (See Asa Jones)

Stephens Axe & Scythe Factory   Oakland
Tools Made: Axes and Scythes

Stetson, Waterman**   Damariscotta   -1855-
Tools Made: Pump and Block Makers
Remarks: He is listed in the 1855 *Maine Registry and Business Directory*.

Stevens & Singer**   Ellsworth   -1856-
Tools Made: Shipsmith
Remarks: They are listed in the 1856 *Maine Business Directory*.

Stevens, E. N.**   Farmington   -1855-
Tools Made: Agricultural Implements
Remarks: He is listed in the 1855 *Maine Registry and Business Directory*.

Stevens, George O.   Unity   -1874-1875-
Tools Made: Edge Tools

Stevens, Grenville M.   Portland   ca. 1872-1873-
Tools Made: Saw Tools (Mitre Boxes)
Remarks: Stevens held two patents (7 May 1872 and 15 July 1873) for a miter box that he made. The mark **STEVENS' MITRE BOARD | PAT'D JULY 20, 1873**, appears on one of his tools; the incorrect patent date was corrected in later marks.

Stevens, John**   Bath   -1881-
Tools Made: Shipsmith
Remarks: He is listed in the 1881 *Maine Business Directory*.

Steward, T. M.**   Anson   1877-1878
Tools Made: Gunsmith
Remarks: Demeritt (1973, 165).

Steward, Williams & Co.   Skowhegan   -1865-1878-
Tools Made: Axes and Edge Tools
Remarks: Owner R. Williams might be Robert. The other owner is S. W. Steward. The name has also been reported as "Steward & Williams." This company is listed in the 1867 and 1869 *Maine Business Directories*. The bill of sale reproduced here, courtesy of Raymond Strout, clearly gives the date of 1865.

Stewart, John B.   Richmond -1849-
Tools Made: Edge Tools

Stiller, C.**   St. John, New Brunswick, Canada   ca. 1840-1850?
Tools Made: Edge Tools
Remarks: See the Canadian toolmakers in Appendix G.

Stimson, J. F.*   Masardis
Tools Made: Axes

Stimson, William*   Springfield   1880-1914
Tools Made: Axes

Stimson, Woodbury   Gray   ca. 1825-1850
Tools Made: Axes and Edge Tools
Remarks: Woodbury used the mark **W. STIMSON | GRAY**, but an 1849 directory listed a T.&W. Stimson.

Stinson, J. F.**   Springfield   -1879-1882-
Tools Made: Axes
Remarks: He is listed in the 1879, 1881, and 1882 *Maine Business Directories*.

Stinson, John F.   Bath   -1869-1871-
Tools Made: Edge Tools
Remarks: He is listed in the 1869 *Maine Business Directory*.

Stinson, R. G.*   Bath   -1874-1878
Tools Made: Axes

Stokes Bros.   Bangor
Tools Made: Chisels
Remarks: A barking spud has been recovered with the mark: **STOKES | BANGOR | BROTHERS**, with quarter moon figures to either side.

Stone, Theodore**   Brunswick   -1795-
Tools Made: Blacksmith

Storer, G. L.**
Tools Made: Planes
Remarks: Various planes have been reported with the **G.L.STORER** imprint, most showing multiple name strikes. Most are made of lignum and are typical of northern New England seacoast boat shop planemaking (Pollak 2001, 397). The Bob Jones collection has a 9 3/8" birch low angle miter plane with three strikes in a triangle and a second plane with two strikes.

Storer, Joshua P.**   Brunswick   1854-1873
Tools Made: Wood Planes
Remarks: Apparently a carpenter and spar maker who made planes, Storer was reported producing 500 planes, made out of both lignum and beech, valued at $750 in 1860. His imprint **J.P.STORER | BRUNSWICK | ME.** with the name line curved are rated as "rare to very rare" according to Pollak (2001, 398). The Liberty Tool Company recently donated a Storer smooth plane made out of lignum vitae to the Davistown Museum. The plane has no wedge or blade but is clearly marked Brunswick, ME. It is in the Maritime IV collection. Two additional examples in the Bob Jones collection are a 9 ½" birch rounding plane and a 9 ½" birch spar plane. Bob St. Peters has a side bead plane marked **J. Storer**. It is unknown if he is related to G. L. Storer.

Strange, J. W.   Bangor   ca. 1860-1880-
Tools Made: Scientific Instruments
Remarks: Listed in directories as a die cutter and machinist, Strange made a navigator's course plotting tool, which combined several functions, such as protractor and T-square, that bore the mark **J.W.STRANGE | Manufacturer | BANGOR, ME ‖ PAT'D June 13, 1876.** This tool was patented by Joseph D. & Samuel Leach of Penobscot, Maine.

Strange held a 11 Sept. 1860 patent for dividers, but whether they were ever made is not known.

Strickland, Asa*   Brighton   -1874-1882-
Tools Made: Axes, Knives, and Shaves
Remarks: He is listed in the 1879, 1881, and 1882 *Maine Business Directories*.

Stubbs, F.**   Sherman   1877
Tools Made: Gunsmith
Remarks: Demeritt (1973, 165).

Swartz, Michael   Bangor   -1849-
Tools Made: Saws
Remarks: This is probably a misspelling of Michael Schwartz.

Sweetser, William**   Bucksport   -1855-
Tools Made: Pump and Block Makers
Remarks: He is listed in the 1855 *Maine Registry and Business Directory*.

Swift, Chauncey**   Montville   -1870-1880-
Tools Made: Blacksmith
Remarks: He is listed in the 1870 and 1880 census.

Swift, W. A.**   Belfast   -1881-
Tools Made: Shipsmith
Remarks: He is listed in the 1881 *Maine Business Directory* on Main St.

Sylvester, Bela P.**   Montville   -1850-
Tools Made: Blacksmith
Remarks: He was born in Freedom, Maine and is listed in the 1850 census.

Tabbutt, A. J.**   Columbia   1875
Tools Made: Gunsmith
Remarks: Demeritt (1973, 165).

Taber, John & Co.   Houlton   -1867-1871-
Tools Made: Axes
Remarks: This company later became John Taber & Sons. It is listed in the 1867 and 1869 *Maine Business Directories*.

Taber, John & Sons   Houlton   -1874-1879-
Tools Made: Axes

Remarks: Followed John Taber & Co. John Tabor (note the spelling difference) of Houlton is listed in the 1879 *Maine Business Directory* as an ax-maker.

Taber, Allen   Augusta   -1870-1880-   (b.1800, d.1882)
Tools Made: Wood Planes
Remarks: Son of Nicholas Taber and brother of John Marshall Taber, both plane makers from New Bedford, MA, he may have worked with them as indicated by his father's will dated 1844, leaving Allen "all of my shop tools" (Pollak 1994). Also, according to the DATM (Nelson 1999), in 1844, Allen moved to Maine, to a farm six miles outside of Augusta and was recorded as a farmer in the 1850 census. He does not appear to have made any planes before moving to a house in town in 1855. He is listed as a planemaker in the 1870 and 1880 censuses, although no examples of his imprint have been reported. In May of 1882 his obituary appeared in the *Kennebec Journal* and said: "He moved to Augusta in 1844 and consequently has been a resident of the city for some 38 years. He was a very quiet, industrious man and highly esteemed by all those who had the pleasure of his acquaintance. Until enfeebled by ill health, he was a manufacturer of bench planes and all carpenters far and near sought for Taber's planes" (Pollak 1994, 403).

Tabor, S. N.**   Houlton   -1881-
Tools Made: Axes
Remarks: He is listed in the 1881 *Maine Business Directory*. He may be related to the Tabers listed above.

Tamminen, Nestor**   Greenwood   1930-1990-
Tools Made: Rules
Remarks: Addison Saunders has a caliper rule, possibly for figuring board feet of logs, that was made by Nestor Tamminen. He was of Finnish heritage and lived in the back country of Maine where he worked as a logger, road-builder, and miner.

Tarbox, F.   Calais   1871-1880
Tools Made: Wood Planes
Remarks: Pollak (2001, 370) notes that "no example of an imprint has been reported."

Tarr, Alex**   Friendship   -1882-
Tools Made: Edge Tools
Remarks: He is listed in the 1882 *Maine Business Directory*.

Taylor, W. & A.**   Portland   -1869-
Tools Made: Shipsmith
Remarks: They are listed in the 1869 *Maine Business Directory* at 50 Commercial St.

Taylor, Johnathan C.**   Bangor   -1855-
Tools Made: Pump and Block Makers
Remarks: He is listed in the 1855 *Maine Registry and Business Directory*.

Taylor, William**   Portland   -1881-1882-
Tools Made: Shipsmith
Remarks: He is listed in the *Maine Business Directory* in 1881 at 46 Commercial St. and in 1882 at 54 Commercial St.

Teague, J. M.   Weld   -1883-
Tools Made: Chisels
Remarks: The DATM (Nelson 1999) notes of the recorded mark: **J.M. TEAGUE | WELD, MAINE || J.M.T. | MAPLE LEAF | 1883**, "the 'MAPLE LEAF' could possibly be a figure vs. the words."

Temple, Levi**   Montville   -1850-
Tools Made: Blacksmith
Remarks: He was born in Montville and is listed in the 1850 census.

Terrill, D. D., Saw Co.*   Bangor   1919-1946
Tools Made: Axes
Remarks: Yeaton (2000) listed the date 1927. Lamond (2007b) gives the dates 1919-1946.

Thaxter, J.**   Portland
Tools Made: Axes

Remarks: The Davistown Museum has an ax (ID# 91303T20) marked **THAXTER PORTLAND CAST STEEL WARRENTED** in the MIV collection (photo). It also has a peen adz (ID# 10606T1) marked **THAXTER PORTLAN_** and a block adz (ID# 40107T2) marked **J. THAXTER PORTLAND CAST STEEL**. No other Thaxter edge tools are known to the editor of this registry. Working dates are unknown. The peen adz appears to be from the 1850 - 70 era and was loaned to the Maine Historical Society for the show "Passionate Pursuits - History in the Collector's Eye," which ran from June 30 - December 2006. Additional information is sought about the identity and working dates of Thaxter, probably a Portland shipsmith who made both edge tools and the iron fittings for locally built sailing ships.

Thomas, Joseph P.**   Blue Hill   -1855-1856-
Tools Made: Edge Tools

Remarks: He is listed in the 1855 and 1856 *Maine Business Directories* as an edge toolmaker.

Thomas, M. B.**   Bucksport   1860-1870-
Tools Made: Gunsmith
Remarks: Demeritt (1973, 165).

Thomas, N.   East Dixfield   ca. 1857-
Tools Made: Household Tools
Remarks: Thomas' stamp: **N.THOMAS | EAST DIXFIELD ME | PAT.OCT 20, 1857** has been found on an apple slicer. The patent is his.

Thomas, Reginald C.**   Cheuncook   1907
Tools Made: Gunsmith
Remarks: Demeritt (1973, 165) notes he had a patent.

Thompson & Merrill**   Augusta   -1855-
Tools Made: Shovels and Spades
Remarks: They are listed in the 1855 *Maine Registry and Business Directory.*

Thompson, Granville**   Montville   -1880-
Tools Made: Blacksmith
Remarks: He is listed in the 1880 census. Also mentioned as being Maimed.

Thompson, H. A.**   Farmington   1880
Tools Made: Wrenches
Remarks: Thompson held the patents on wrenches produced by the Diamond Wrench Co. of Portland. Herb Page owns the 12" long combination nut and buggy wrench (left photo); **November 2, 1880 patent** is cast into the lower jaw. The Davistown Museum has a 5 3/8" long double jaw buggy wrench (ID# 041403T1) in the IR collection (right photo).

Thompson, H. P.   Belfast   -1862-
Tools Made: Household Tools (Spice and Coffee Mills)

Thompson, M. N.**   Anson,   -1856-
Tools Made: Edge Tools
Remarks: He is listed in the 1856 *Maine Business Directory.*

Tibbets, George H.**   Augusta   1872
Tools Made: Gunsmith

Remarks: Demeritt (1973, 165) notes he had a patent.

Tilton, W., Jr.** Livermore -1855-
Tools Made: Agricultural Implements
Remarks: He is listed in the 1855 *Maine Registry and Business Directory.*

Titus, Daniel Butters East Union -1850-1910- (b.1829)
Tools Made: Coopers' Tools, Coopers' Planes

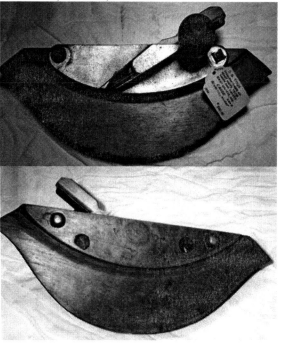

Remarks: Born in Union, Daniel, whose name has also been reported as Danial (Nelson 1999), was noted as a carpenter in 1850 and a maker of mast and truss hoops and coopers' tools in 1880. A coopers' croze, bearing a paper label and a circular mark reading: **D.B.TITUS | Manufacturer of | MAST HOOPS, TRUSS | HOOPS AND CROZES | EAST UNION,ME** is in the Bob Jones collection with the stamp of **H.H.CRIE & CO.**, a hardware dealer in Rockland, Maine, along its top edge (Pollak 2001, 413). DATM (Nelson 1999) notes a second mark: **D.B.TITUS | EAST UNION.** Additional information on Titus, who must have supplied tools to the nearby cooperages of Liberty, is sought by The Davistown Museum.

Tobey, Lemuel** Portland 1823
Tools Made: Gunsmith
Remarks: Demeritt (1973, 165).

Tobey, Lemuel, Jr.** Portland 1823
Tools Made: Gunsmith
Remarks: Demeritt (1973, 165).

Todd, B.** Albany -1820-
Tools Made: Planes
Remarks: Pollak (2001, 414) states this is possibly B. P. Todd who is listed as a cabinetmaker in Albany, ME. An example is a 9 9/16" beech, 1 3/16" ogee molder with heavy round chamfers, marked **B.TODD.**

Torrey & Holmes**   Rockland   -1855-
Tools Made: Brass Founders and Finishers
Remarks: They are listed in the 1855 *Maine Registry and Business Directory*.

Torrey, David**   Deering   1873
Tools Made: Tool Holder
Remarks: He has Pat. No. 136,450, for an improvement in tool holder, Mar. 4, 1873. One half is assigned to Ira F. Munch of Sumner, ME (Tom Lamond, personal communication).

Torrey, F. B.**   Richmond   -1855-
Tools Made: Brass Founders and Finishers
Remarks: He is listed in the 1855 *Maine Registry and Business Directory*.

Towle, Levi & J. M.*   Westbrook   1809-1856
Tools Made: Axes

Trafton, Charles**   York   1706
Tools Made: Gunsmith
Remarks: Demeritt (1973, 6, 165).

Trafton, Joseph**   Machiasport   -1855-
Tools Made: Wheelwright
Remarks: He is listed in the 1855 *Maine Registry and Business Directory*.

Treat, John**   Thomaston   1749-1759
Tools Made: Gunsmith
Remarks: Demeritt (1973, 165).

Treat, Joshua**   Fort Pownal, Thomaston, Stockton Springs   1750-1774
Tools Made: Gunsmith and Armorer
Remarks: He is listed as the first English settler on the Penobscot River. He served as gunsmith and armorer for both George's Fort and Fort Pownal. He understood the Penobscot Indian language and accompanied Governor Pownal on his trip up the Penobscot River in 1759 (*Biographical Review* 1897). Demeritt (1973, 11, 165) notes his other working locations and the date range of 1759-1774.

Tripp, S.**   Rockland   -1881-
Tools Made: Shipsmith
Remarks: He is listed in the 1881 *Maine Business Directory* on Main St.

Trott & Lemont*   Bath   1835-1850-
Tools Made: Hoops, Anvils, Vises, Crowbars, Plough Molds, Mill Saws, Fishermen's Anchors, and Edge Tools
Remarks: Alfred Lemont and James F. Trott partnership (Baker 1973, 429).

Trott, James F.*   Bath   -1835-
Tools Made: Blacksmith
Remarks: A master blacksmith who for awhile partnered with Alfred Lemont. See Trott & Lemont.

Trott, J. S.*   Castine   -1855-
Tools Made: Wheelwright
Remarks: He is listed in the 1855 *Maine Registry and Business Directory*.

True, J. L.   Benton   -1871-
Tools Made: Farm Tools (Potato Planters)

True, Lyman C.*   Pownal   -1881-1882-
Tools Made: Axes, Ice Tools, and Knives
Remarks: He is listed in the 1881 and 1882 *Maine Business Directories*.

True, W. M.   Waterville   -1899-1900-
Tools Made: Farm Tools

Tufts (See Asa Jones)

Tuller, Leonard   Farmingdale   -1855-
Tools Made: Edge Tools

Turner, Charles*   Prospect   ca. 1800
Tools Made: Blacksmith
Remarks: Turner built and worked at the blacksmith shop that in the 1890s, was referred to as "the old blacksmith shop behind the K of P hall." John Libby apprenticed under Charles Turner (Ellis 1980, 251-2).

Turney, R. L.   Houlton   -1899-1900-
Tools Made: Farm Tools

Tuttle, Rufus*   South Durham, Durham   -1855-1879-
Tools Made: Axes
Remarks: He is listed in the 1855, 1856 and 1879 *Maine Business Directories*.

Tylor, George W.**   Lowell   -1856-
Tools Made: Edge Tools
Remarks: He is listed in the 1856 *Maine Business Directory*.

Ullrich Bros.   Caribou   -1899-1900-
Tools Made: Farm Tools

Union Axe Company**   Oakland   1904-1925-
Tools Made: Axes
Remarks: Klenman (1990, 26) lists Union Axe company as one of the Oakland ax-makers. Klenman (1990, 26) reprints their paper label for their fire ax, which also notes "manufactured for N.Y. Belting & Packing CO Chicago, Ill." Judging from the depiction of the firemen, Union Axe Company was probably an early 20th century manufacturer. Lamond (2007b) gives the dates 1904-1925.

Vannah Co.   Gardiner   -1899-1900-
Tools Made: Farm Tools

Varney, Thomas & Son   Windham   -1855-
Tools Made: Farm Tools

Varney, Aaron & John B.   North Berwick   -1855-1862-
Tools Made: Farm Tools and Plows
Remarks: Aaron was listed alone in 1855 as a maker of farm tools, but, then, in 1862, both he and John B. Varney are listed, albeit separately, as makers of "plows and all kinds of castings" in North Berwick. It is assumed that they worked together (Nelson 1999).

Varney, George   Kennebunk   -1885-
Tools Made: Plows

Vaughan, Daniel**   New Vineyard   -1855-1856-
Tools Made: Edge Tools
Remarks: He is listed in the 1855 and 1856 *Maine Business Directories*.

Vaughan, J and Co. **   Union   1850-1856-
Tools Made: Shovels and Spades
Remarks: They are listed in the 1855 and 1856 *Maine Registry and Business Directory* as Vaughn & Co. See the Sibley quote under "Vaughan & Pardoe."

Vaughn & Cobb**   Union

Tools Made: Slicks

Remarks: A 4" slick marked **Vaughn & Cobb | Union | Warranted** has been reported. The owner has found references to a Cobb having lived in Union, Maine, but no references to any as toolmakers.

Vaughan & Pardoe*   Union   1844-1868

Tools Made: Axes, Edge Tools, and Coopers' Tools

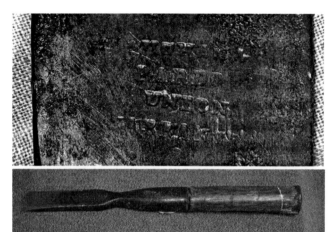

Remarks: The Davistown Museum collection includes Vaughan & Pardoe gouges (ID#s 31501T5, 31908T16, and 111001T3 [photo of mark]), a gutter adz (ID# 61204T2), slicks (ID#s 062603T2 and 21201T3 [photo]), a blubber cutter (ID# TKD1301), and a socket chisel (ID# 41801T7). One slick is marked **VAUGHAN PARDOE & Co UNION WARRANTED.** Other tools are marked **VAUGHAN & PARDOE UNION WARRANTED.** DATM (Kijowski 1990) lists Vaughn and Pardoe as a Union, Maine, maker of drawknives with no date, but there is no listing in DATM (Nelson 1999). The dates we list come from Yeaton (2000). Sibley (1851) writes "IRON WORKS. -- In June, 1843, an iron-foundery was established at South Union. Here 'are made all kinds of country castings.' In August, 1844, business was commenced in the edge-tool factory of Vaughan and Pardoe. Nearly four thousand axes are made annually; also ship-tools to the value of about $1,500, and coopers' tools to about the same amount. March 12, 1850, **J. Vaughan and Co.** commenced business in their shovel-factory. The manufactures at all these establishments are regarded as of a very superior quality; as well as the tool-work of Bradley R. Mowry, at the Middle Bridge." Vaughn and Pardoe's edge tool production coincides with the peak period of shipbuilding in the nearby Waldoboro Customs District.

Vaughan, Pardoe & Cox**   Union

Remarks: The Museum has a coopers' adz (ID# 12801T7) marked **VAUGHAN | PARDOE & COX | UNION | WARRANTED** in the MIII collection, the only tool we know of with this signature. See "Vaughan & Pardoe."

Venner, O. H.**   Rockland   1877

Tools Made: Gunsmith

Remarks: Demeritt (1973, 165).

Vloom, Wm.**   Calais   -1856-
Tools Made: Shipsmith
Remarks: He is listed in the 1856 *Maine Business Directory.*

Vupum, James & Co. **   Gardiner   -1855-
Tools Made: Pump and Block Makers
Remarks: He is listed in the 1855 *Maine Registry and Business Directory.*

W., A. (See A. Welles)

W., J. H.**   Waldoboro   ca. 1850
Tools Made: Planes
Remarks: Pollak (2001, 219) states that the mark **J.H.W. | WALDOBORO** was possibly made by someone related to Thomas Waterman of Waldoboro, Maine. His example is a 9 ½" beech skew rabbet in the Bob Jones collection.

W. S. & A. Co.   Gardiner
Tools Made: Wrenches

Waldron, Alexander**   Kittery   1660-1676
Tools Made: Gunsmith
Remarks: Demeritt (1973, 165).

Walker, Abiel F.**   Alna   b. 1808 - d. 1875
Tools Made: Planes

Remarks: "Abiel F. Walker was a carpenter/cabinetmaker living in Alna, Maine (he may also have worked in a boatyard in Bath). He was active ca. 1830 – 1860's. He, or a family member, also had skills as a smith and hence forged some of his own tools such as plane blades" (David Walker - great great grandson, 9/19/01). The Davistown Museum obtained a collection of Abiel's planes from the attic of the Alna house where Abiel lived (ID# 42602T1, photo). The house was built in 1783. Large areas of Alna and Wiscasset were owned by the Walker clan in the 19th century. Abiel participated in building many wooden structures and boats in the Wiscasset area in the boomtown years of Maine's maritime era. He signed his planes only with his initials **AFW**. He made sufficient numbers of planes that he may have supplied neighboring carpenters and boat builders. See the essay describing the context of Abiel Walker's planemaking in Appendix E. Also see *Walker Family History* (American Genealogical Research Institute 1972).

Walker, James, Jr.**   Belfast   1860-1870-
Tools Made: Gunsmith

Remarks: Demeritt (1973, 165).

Walker, John*   Scarborough   1832-1863
Tools Made: Axes

Walker, R.**   -1820-
Tools Made: Planes
Remarks: Pollak (2001, 426) lists a 9" beech #18 round with round chamfers and a relieve wedge marked **R.WALKER** found in Maine.

Walker, S. W.   Anson   -1871-
Tools Made: Rakes (Horse Rakes)

Wallace, G. E.**   Jackson
Tools Made: Planes
Remarks: The mark **G.E.WALLACE | JACKSON.ME.** was found on a 9 3/8" long beech complex molder. It was reported in the *Sign of the JOINTER,* Volume 4, Number 2, Fall 2002.

Walton, S. B.**   East Livermore   -1881-1882-
Tools Made: Edge Tools
Remarks: He is listed in the 1881 and 1882 *Maine Business Directories.*

Waltz, D. O.**   ?   -1830-
Tools Made: Planes
Remarks: Pollak (2001, 427) lists a 9 3/16" beech stair rail plane with a bench-style wedge, a single iron, shallow round chamfers, marked **D.O.WALTZ** found in Maine.

Waltz, William G.**   Waldoborough   -1869-
Tools Made: Shipsmith
Remarks: He is listed in the 1869 *Maine Business Directory.*

Warner, N. E.   Wilton   -1879-
Tools Made: Farm Tools

Washborn, Thomas**   Richmond   1740-1742
Tools Made: Gunsmith
Remarks: Demeritt (1973, 165).

Washburn, Oscar   Portland   ca. 1840-1845-
Tools Made: Wood Planes
Remarks: Oscar Washburn was known as a planemaker in Goshen, MA, around 1845

and may possibly have been involved with the Union Tool Co., also of Goshen (Nelson 1999). A circa 1840 grooving plane bearing the imprint **O:WASHBURN | PORTLAND** has been recovered (Pollak 2001, 432). The Bob Jones collection has a dado plane marked **O:WASHBURN** with no location noted.

Wass, Rufus**   Addison   -1855-
  Tools Made: Wheelwright
  Remarks: He is listed in the 1855 *Maine Registry and Business Directory*.

Waterhouse, Emery & Co.**   Portland   1842-2000-
  Tools Made: Planes
  Remarks: The Bob Jones collection has a 9 ½" birch skewed rabbet plane with slightly rounded chamfers marked **EMERYWATERHOUSE&Co. | PORTLAND,ME**. The company name lettering is arced in a semi-circle. Lamond (2007b) states this company is a hardware distributor using proprietary brands.

Waterhouse, W. H.   Gardiner   -1869-
  Tools Made: Axes
  Remarks: He used the mark: **WATERHOUSE/GARDINER ME**. See Aldrich & Waterhouse.

Waterman, Thomas   Waldoboro   1800-1850   (b. ca. 1775)
  Tools Made: Wood Planes
  Remarks: Pollak (1994, 398) states that Thomas, born around 1775, was the son of Abijah Waterman, a merchant in the town of "Waldoborough," which, at that time, was a province of Massachusetts. His father died in 1782, and, on October 7, 1794, Thomas "a minor above the age of 14 years" was made the ward of Charles Samson. Pollak also states that "His planes are 9 ½" long, birch, have relieved wedges, broad flat chamfers, and appear to be professionally made" and are extremely rare with about 10 to 50 examples estimated.

Pollak's (2001, 433) 4th edition has removed some of the previous comments but adds: "In 1809 he was surveyor of lumber for Waldoboro, and was a selectman in 1819." His planes were imprinted **T*WATERMAN | WALDOBORO**. The Davistown Museum has three Waterman planes in the MII collection (ID#s TBW1002, TBW1003, and TBW1004) that are made of beech rather than birch, the former being the apparent wood of choice for most of Waterman's planes and the birch possibly being the wood used in the oldest of Waterman's planes. The fall 2002 issue of *Sign of the JOINTER* reports two more T. Waterman planes, a birch 9 ¾" rounding plane and a 9 7/8" birch hollow plane.

Waterman is considered one of Maine's earliest planemakers. See the essay on him in the "Maine's Earliest Planemakers" chapter. The two Waterman planes (photo) are in the Bob Jones collection. The top is a 10 7/8" birch closed tote ships hollow with steel wear strips on both sides. The bottom is a 10" birch open tote gutter plane with finder rails on one side. His collection also has a 9 ½" quarter round moulding plane (not pictured.)

Watson, John   Houlton   -1899-1900-
Tools Made: Farm Tools

Waugh, Joseph**   Levant   1875
Tools Made: Gunsmith
Remarks: Demeritt (1973, 165).

Waugh, Lewis**   Levant   1875
Tools Made: Gunsmith
Remarks: Demeritt (1973, 165).

Waugh, Th.**   Somerset County   1818
Tools Made: Blacksmith
Remarks: He is listed in the Somerset County Probate Inventories.

Waugh, Thomas**   Starks   1871
Tools Made: Gunsmith
Remarks: Demeritt (1973, 166).

Wayne Scythe Mfg. Co.   Wayne   1838-1845
Tools Made: Scythes
Remarks: The owner of this company from 1840, Reuben Dunn, ran the business until he sold it in 1845 to what became the North Wayne Scythe Co.

Wayne Tool Co.   Hallowell   ca. 1889-
Tools Made: Farm Tools
Remarks: A hay knife/saw that was patented by Marcus M. Bartlett of Hallowell was manufactured by this company, which may be the same as the North Wayne Tool Co. The mark found was **WAYNE CO. MFRGS | PAT.----89**.

Webb & Nason   Bangor   -1875-1887
Tools Made: Axes, Carpenter Tools, Coopers' Tools, and Shaves
Remarks: This company is listed in the 1879, 1881, and 1882 *Maine Business Directories*. See Lester Webb.

Webb, F. A. Bridgton -1899-1900-
Tools Made: Farm Tools

Webb, Lester Bangor ca. 1875
Tools Made: Axes and Drawknives
Remarks: Two marks that have been identified are **L.WEBB | BANGOR** and **L.WEBB SUCCESSOR | TO HIGGINS & WEBB**. The Davistown Museum has an L. Webb drawknife (ID# TJC1001) in the IR collection. Also see Higgins & Webb.

Webber, Haviland & Philbrick** Waterville -1895-
Tools Made: Planers, Saw Mills, and Iron Founder
Remarks: Bill of sale courtesy of Raymond Strout.

ESTABLISHED IN 1853.
Terms _____ Waterville, Me., *May 18th* 18*75*
Messrs S Willsley.
Bought of WEBBER & HAVILAND,
WEBBER, HAVILAND & PHILBRICK,
Machinists, Brass and Iron Founders,
MANUFACTURERS OF
SINGLE, GANG, CIRCULAR AND MULEY SAW MILLS, WOODWORTHS & DANIELS' PLANERS,
ALSO, E. TUTTLE'S PATENT SELF REGULATING WATER WHEEL.

Weeks & Whitney Sebago Lake
Tools Made: Wrenches

Welch, P.** Calais -1856-
Tools Made: Shipsmith
Remarks: He is listed in the 1856 *Maine Business Directory*.

Weld, Moses Greenbush, Olamon -1867-1889-
Tools Made: Axes and Chisels
Remarks: Weld used both Greenbush and a neighboring town, Olamon, for his address. He used the imprint **MOSES WELD | OLAMON ME.** on his axes. He is listed in the 1867, 1869, 1882, and 1879 *Maine Business Directories*.

Welles, A.** ? -1850-
Tools Made: Planes
Remarks: Pollak (2001, 437) lists a 16" beech jack with a centered open tote and a single Barry & Way iron marked **A.WELLES** and an **A.W.** found in Maine.

Wells, John C.** North Bridgton 1891-
Tools Made: Gunsmith
Remarks: Demeritt (1973, 166).

Wentworth, H. A. Greene -1899-1900-
Tools Made: Farm Tools

Wescott, George**   Prospect   -1872-

Tools Made: Blacksmith

Remarks: Wescott and George Avery owned and ran Bowdoin Point Blacksmith. The hill upon which the blacksmith shop sat is still referred to as Blacksmith Shop Hill. They were connected with the quarries (Ellis 1980, 251-2).

Wescott, H. W.   Knox   -1869-

Tools Made: Farm Tools

Remarks: Wescott made horseshoes; it is not known if he made other tools as well.

West, D. S.**   Auburn   1894

Tools Made: Gunsmith

Remarks: Demeritt (1973, 166).

West, J. C.**   Letter B   -1855-

Tools Made: Edge Tools

Remarks: He is listed in the 1855 *Maine Business Directory*. The town is given as "Letter B," perhaps this means that, instead of a town name, only the letter B was written down.

West, Lyman M.**   Franklin   -1855-

Tools Made: Wheelwright

Remarks: He is listed in the 1855 *Maine Registry and Business Directory*.

Weston, Charles H.   Yarmouthville   ca. 1858-1875-

Tools Made: Shaves

Remarks: Weston moved to Maine shortly after patenting a spoke shave in Nashua, NH. There are two distinct marks he used. **C.H.WESTON | PAT'D JUNE 1, 1858** has been reported both as simply linear or as a circular mark. The other mark plainly reads **WESTON**.

Wethen, D. Y.   Detroit   -1849-

Tools Made: Edge Tools

Wetherall, Samuel   Whitefield   1880-

Tools Made: Edge Tools?

Remarks: Wetherall was a blacksmith associated with the Kings Mills complex in Whitefield and almost certainly made tools for the shipwrights of the Boothbay region.

Wetherbee, Timothy**   Bath   -1882-
Tools Made: Shipsmith
Remarks: He is listed in the 1882 *Maine Business Directory* on Water St.

Wetherell (see Witherell)

Weymouth*   Belgrade   1879-1890
Tools Made: Axes

Weymouth, George F.**   Dresden   -1879-
Tools Made: Hay Knives
Remarks: He is listed in the 1879 *Maine Business Directory*. He patented a hay knife, patent number 112,400 March 7, 1871. This patent was sold to Hiram Holt & Co. sometime in the 1870s.

Weynouth, John & Son   Sangerville   -1885-
Tools Made: Farm Tools

Whalen, James**   Lubec   -1856-
Tools Made: Shipsmith
Remarks: He is listed in the 1856 *Maine Business Directory*.

Wheeler & Lawrence**   Farmington
Tools Made: Gunsmith
Remarks: Demeritt (1973, 83-4, 166) notes this is Albert Galletin Wheeler and William Lawrence.

Wheeler & Stevens**   Farmington   1872
Tools Made: Gunsmith
Remarks: Demeritt (1973, 83-4, 166) notes this is Albert Galletin Wheeler.

Wheeler, Albert Galletin**   Farmington   1856-1883   (d. 1883)
Tools Made: Gunsmith
Remarks: Demeritt (1973, 83-4, 166) notes he worked as A. G. Wheeler & Co., Wheeler & Stevens, and Wheeler & Lawrence.

Wheeler, Charles E.**   Farmington   1871-1900   (d. 1916)
Tools Made: Gunsmith and Fishing Rods
Remarks: Demeritt (1973, 166) notes he is A. G. Wheeler's son.

Wheeler, Joel B.**   Etna   1877

Tools Made: Gunsmith
Remarks: Demeritt (1973, 166).

Wheeler, P.**   ?   -1790-
Tools Made: Planes
Remarks: Pollak (2001, 441) lists a 14 5/8" gutter plane with a double-pegged tote and a 9 1/2" fenced skew rabbet, both birch with heavy flat chamfers, and a birch tongue plane all marked **P.WHEELER** found in Maine. Two other examples were found in Fitchburg, MA.

Wheeler, S. A. & Son**   Waterville   1874-1883
Tools Made: Gunsmith
Remarks: Demeritt (1973, 166).

Wheelock & Ames**   Portland   1832
Tools Made: Gunsmith
Remarks: Demeritt (1973, 166).

Whepley, Henry**   Eastport   -1856-
Tools Made: Shipsmith
Remarks: He is listed in the 1856 *Maine Business Directory*.

Whipple, Carlyle   Lewiston   ca. 1857-
Tools Made: Saws
Remarks: Whipple held a patent with the date of 18 Jan. 1857 for a "reciprocating saw mill" that he made.

White, C.   Auburn   -1869-
Tools Made: Brushes

White C.**   Monroe   -1855-1856-
Tools Made: Edge Tools
Remarks: He is listed in the 1855 and 1856 *Maine Business Directories*.

White, Chandler   Dixmont, Bangor   -1849-1871-
Tools Made: Edge Tools
Remarks: In 1849, White was located in Dixmont, Maine, but, by 1869, he had moved to Bangor and is listed there in the *Maine Business Directory*.

White, D. & Sons   Portland   -1869-
Tools Made: Brushes

White, Ephraim R.**   Waldoboro   -1860-1900-   (d. 1933)
Tools Made: Axes, Edge Tools, and Gunsmith
Remarks: Ephraim White is listed in the 1881 and 1882 *Maine Business Directories* for making axes and edge tools. Demeritt (1973, 166) lists him as making half stock percussion rifles and pistols.

White, George E.**   Belfast   -1881-
Tools Made: Shipsmith
Remarks: He is listed in the 1881 *Maine Business Directory*.

White, James**   South Portland   2001-
Tools Made: Planes
Remarks: In 2001, he purchased the Crown Plane Co. from Leon Robbins of Bath. He uses the mark **JW** with a crown emblem. The Davistown Museum has a plane he made for the Museum in the IR collection (ID# 90502T1). See the Crown Plane Co. listing for a photograph of the plane.

White, Job**   Belfast   -1829-
Tools Made: Carriage-maker
Remarks: Job White and Phineas Quimby received an 1829 patent for a "machine for cutting veneers and panels for carriages". Information courtesy of Jeff Joslin.

White, P.**   Jackson   -1855-1856-
Tools Made: Edge Tools
Remarks: He is listed in the 1855 and 1856 *Maine Business Directories* as an edge toolmaker.

White, William   Waldoboro   1856-1871-
Tools Made: Axes and Edge Tools
Remarks: He is listed in the 1856, 1867, and 1869 *Maine Business Directories*.

Whitehouse, James & N.   Smithfield   -1869-
Tools Made: Farm Tools
Remarks: The loose assumption that James and N. Whitehouse may have worked together is based on the location and an 1869 directory which listed both Whitehouses, albeit separately, for similar items. James made agricultural implements and N. made harrows and carts.

Whiting, J. R. & Co.   Belfast   1834-1856-
Tools Made: Axes and Edge Tools
Remarks: Whiting used the mark:
WHITING | BELFAST. John R.
Whiting is listed in the 1855 *Maine Business Directory* as an edge toolmaker. A copy of the *Republican Journal* (Nov. 11, 1834) contained this advertisement.

Whiting, Steven*   Belfast
Tools Made: Axes

Whitman Agricultural Works
Auburn   -1899-1900-
Tools Made: Farm Tools

Whitman, Luther   Winthrop   -1855-1871-
Tools Made: Farm Tools and Plows

Whitman, W. E.   Winthrop   -1879-1885-
Tools Made: Farm Tools
Remarks: He is possibly related to Luther Whitman.

NEW EDGE-TOOL FACTORY.
JOHN R. WHITING & Co

HAVE commenced and intend to carry on the business of

EDGE-TOOL MAKING,

in all its various branches at their Edge-Tool Factory near Gannett's mills, in Belfast.—Those who may favour them with their paronage may rest assured that no pains will be spared by experienced workmen to make their tools equal if not superior to tools made in other States. All tools made at the "Belfast Edge Tool Factory" will be *warranted* 30 days from the date of purchase, and if they should prove bad, they may be returned, and new ones will qe given in exchange.

If any person should like axes or other tools made in any particular shape, the company would be pleased to have them furnish a pattern, and they will make tools to fit them.—The patronage of traders is respectfully solicited.

OLD MADE NEW.

JOHN R. WHITING & Co., will new-steel and grind tools of every discription, and warrant them good, (equal in every raspect to new.)—Tools for new laying may be left with S. A. Moulton, main-st, or with B. G. *Whiting*, No. 11, Phœnix Row, and mey will be returned to the same places when done.
Belfast, Nov. 11, 1834.   40tf

Whitney Bros.**   Lewiston   1887
Tools Made: Gunsmith
Remarks: Demeritt (1973, 166) notes this is F. R. Whitney and H. A. Whitney.

Whitney, B.*   Belfast
Tools Made: Axes

Whitney, Micah**   Gorham, Gray, Phillips   (b.1752 d.1832)
Tools Made: Blacksmith
Remarks: Micah was born in York, Maine, son of Abel. He served in the Revolutionary War, having enlisted in Buxton. Source: Mrs. Charles Toothaker of Phillips.

Whittemore, Ebeneazor**   East Livermore   (d. 1855)
Tools Made: Blacksmith
Remarks: Source: Androscogin County, Maine Probate Court Records.

Whittemore, James M.**   Augusta   1869-1876
Tools Made: Gunsmith
Remarks: Demeritt (1973, 166) notes he was a U. S. Army patentee at the Kennebec Arsenal in Augusta.

Whitten, Ivory**   Belmont   -1879-
Tools Made: Coopers' Tools
Remarks: He is listed in the 1879 *Maine Business Directory*.

Whitten, Samuel**   Montville   -1860-
Tools Made: Blacksmith
Remarks: He is listed in the 1860 census.

Whorff & Cleveland**   Skowhegan   -1879-
Tools Made: Axes
Remarks: This company is listed in the 1879 *Maine Business Directory*. Note below, under "Barnett Whorff," that DATM (Nelson 1999) states Barnett Whorff began an ax company with S.F. Cleveland in 1861. This may or may not be the same business listed only under Barnett Whorff in 1867.

Whorff, Barnett   Madison, Skowhegan   1855-1888-
Tools Made: Axes
Remarks: Barnett started working with his father, John, in Madison. Between the years 1855-61 they worked together. He is listed in the 1855 *Maine Business Directory* as Barnet Whorff. By 1862, Barnett was listed by himself in Madison, however, a year earlier (1861) he had bought a machine shop in Skowhegan and with one of the shop's previous partners, S.F. Cleveland, began an ax business. The business' first name is unknown, but by 1867 it was recorded under Barnett Whorff's name alone. His father went to work for him, continuing until at least 1872. This business is listed in the 1867, 1869, and 1882 *Maine Business Directory* as located in Skowhegan. It is listed in the 1881 *Maine Business Directory* as Whorff Barnet & Co.

Whorff, James   Madison   -1855-
Tools Made: Axes
Remarks: He is listed in the 1855 *Maine Business Directory*. See John Whorff Sons.

Whorff, John   Madison   ca. 1804-1855   (b. 1784, d. 17 May 1876)

Tools Made: Axes

Remarks: Besides working in Madison, John also worked in Bath and Norridgewock, although it is not clear the years in which he was in those places. He was established as 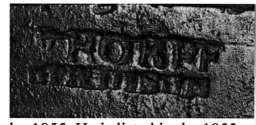 John Whorff & Sons in Madison by 1855. He is listed in the 1855 *Maine Business Directory*. His son, Barnett, began his own company in Skowhegan, and John continued to work with him. The mark **WHORFF 1829** has been found on an ax; a short note by Brundage (1989) on this Whorff ax can be found in *The Chronicle*. The Davistown Museum has a large hewing ax (ID# 21201T2) marked **WHORFF | MADISON** and donated by Rick Floyd in the MIII collection (photo). Also see the Davistown Museum website information file on Whorff.

Whorff, John & Son   Madison   ca. 1855-1861

Tools Made: Axes and Chisels

Remarks: In 1855, John and his two sons, Barnett and James were listed as working together in Madison, but no specific name for this group was recorded until 1860 when John Whorff & Son was recorded. The "son" is assumed to be Barnett. It was around 1861 when Barnett bought a new shop that they moved to Skowhegan, where at least John and Barnett continued to work together under Barnett's name. A barking spud marked **J. WHORFF SONS** has been reported.

Whyley, Luther**   Portland   1846

Tools Made: Gunsmith

Remarks: Demeritt (1973, 166).

Whyley, John, Jr.**   Portland   1831-1863

Tools Made: Gunsmith

Remarks: Demeritt (1973, 166) notes his name is also spelled Wiley.

Wilder, John W. & Son**   Belfast   -1855-

Tools Made: Plow Maker and Agricultural Implements

Remarks: They are listed in the 1855 *Maine Registry and Business Directory*.

Wiley, John, Jr.

Remarks: See John Whyley, Jr.

Willey, Enoch B.**   Cherryfield   1856
Tools Made: Gunsmith
Remarks: Demeritt (1973, 166).

William, Web & Son**   Warren   -1855-
Tools Made: Brass Founders and Finishers
Remarks: They are listed in the 1855 *Maine Registry and Business Directory.*

Williams & Howard**   Bath   -1881-
Tools Made: Shipsmith
Remarks: They are listed in the 1881 *Maine Business Directory* on Commercial St.

Williams & Morse   Skowhegan   1857-
Tools Made: Chisels
Remarks: Shortly before 1857, three brothers, Charles A., Frank A., and Robert Williams, moved to Skowhegan. One of them was the chief partner in this company. During the same time and at the same place, Frank A. ran a foundry, and Charles A. was noted as a chisel-maker.

Williams, Charles A.   Skowhegan   ca. 1860-1900
Tools Made: Chisels and Hatchets
Remarks: Charles' edge tool production evolved out of the Williams & Morse factory. It is not clear under what name he continued to work. Whether he or his brothers had any connection with Steward Williams & Co. of Skowhegan is not known. C. A. Williams is listed as a chisel-maker in the 1867 *Maine Business Directory*. The business is listed in the 1881 and 1882 *Maine Business Directories* as Charles A. Williams & Co. makers of lath hatchets. The Davistown Museum has a lathing hatchet (ID# 43006T9) marked **C. A. WILLIAMS & CO** and a one inch timber framer's socket chisel (ID# 113004T1) marked **C. A. WILLIAMS & CO** and **CAST STEEL** in the IR collection.

Williams, Steward & Co.   Skowhegan   -1870-1871-
Tools Made: Axes
Remarks: See Charles A. Williams.

Wilson**   Lewiston
Tools Made: Drawshaves
Remarks: The Davistown Museum has a drawshave (ID# 111001T13) marked **Wilson Lewiston** with an 8 point asterisk touchmark and an early 19th century appearance in the MIII collection.

Wilson, Bradford**   Acton   -1879-1882-
Tools Made: Knives (Shoe, Butcher, and Jack)
Remarks: He is listed in the 1879, 1881, and 1882 *Maine Business Directories*.

Wilson, E.**   Monmouth   1871
Tools Made: Gunsmith
Remarks: Demeritt (1973, 166).

Wilson, Frank R.**   Houlton   1899
Tools Made: Gunsmith
Remarks: Demeritt (1973, 166).

Wilson, J. S.**   Solon   1880-1881
Tools Made: Smith
Remarks: He is listed in Hoyt (1881).

Wing, Calvin**   Gardner   1830-
Tools Made: Reaction Water Wheel
Remarks: Calvin Wing was the inventor and manufacturer of the reaction water wheel, predecessor to the turbine, which he patented in 1830. The reaction wheel was a modification of the tub wheel. Using a cast iron tub with a large opening on one side and six smaller holes around its perimeter, the reaction wheel used the pressure of the water being forced into the wheel by gravity and out of the wheel through the perimeter openings to turn the wheel. It was more durable than a wooden wheel, but was also more complicated and difficult to install and operate. The reaction wheel would lead to the invention of the turbine. (Sturbridge Village 2008, http://www.osv.org/explore_learn/waterpower/reaction.html).

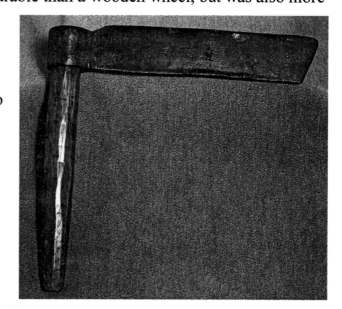

Winn, W M**   Clinton
Tools Made: Froe
Remarks: The Davistown Museum has a froe (ID# 032103T1) made by W. M. Winn in the MIV collection (photo).

Winslow, James H.   Thomaston   - 1842-1865-   (b.1817)
Tools Made: Wood Planes
Remarks: He was born in Nobleboro, Maine, in 1817, and was listed as working there

in 1840. James married in 1842 and moved to Thomaston as a carpenter (Pollak 2001, 453). He worked there from 1842 to 1865, doing carpentry and making planes which were imprinted: **J.H.WINSLOW | THOMASTON | ME**. The pictured 18 ¾" birch razee cornice plane is in the Bob Jones collection. This plane has an unusual blade imprint: **DWIGHTS FRENCH & CO | WARRANTED | CAST STEEL**.

Winslow, Thomas**   Portland   -1869-
Tools Made: Shipsmith
Remarks: He is listed in the 1869 *Maine Business Directory* at 277 Commercial St.

Winslow, William A.   Bath, Winslow   -1867-1880-
Tools Made: Edge Tools
Remarks: Rick Floyd has a socket chisel (photos) marked **WILLIAM | WINSLOW | CAST STEEL**. He is listed in the 1867 and 1869 *Maine Business Directories* as working in Bath.

Winslow, Brown & Co.**
Portland   -1855-
Tools Made: Brass Founders and Finishers
Remarks: They are listed in the 1855 *Maine Registry and Business Directory*.

Witham, Ephraim   Carratunk Plantation   -1860-1884-
Tools Made: Axes, Cant Dogs, and Hammers
Remarks: According to one directory, he was a blacksmith and maker of "driving tools" (sledges and wedges). His name has been incorrectly spelled in another as "Wisham." In the later years, it is assumed that his son joined him, as he added "& Son" to his name. He is listed as making axes in the 1867 and 1869 *Maine Business Directory*. E. Witham & Son are listed in Hoyt (1875) and Hoyt (1881).

Witham, Claude L.*   Carratunk   1900-1929
Tools Made: Axes

Witham, Gustave & Son*   Carratunk   1884-1929
Tools Made: Axes

Witherell, J. H.**   Oakland   1885-1916?
Tools Made: Axes

Remarks: The axes made by J. H. Witherell may have been made at the Witherell Scythe Co., as implied by Klenman (1997). At the least, the J. H. Witherell company was part of that Oakland, Maine, Emerson Stream ax-makers complex, which was the location of many other Maine ax-makers and tool companies, such as the famous Emerson and Stevens Manufacturing Company. Klenman (1990, 26) lists 14 other Oakland ax-makers and reproduces a Witherell ax label, date unknown, with the notation in the lower right corner that the Pierce Hardware Company of Taunton, Mass. was one of their vendors.

Witherell, Samuel B.   Norridgewock   -1862-
Tools Made: Blacksmith and Farm Tools
Remarks: Although he was a blacksmith and dealer in harrows and plows, it is not known if Witherell was actually a maker of either. For a time, he was in business with S. M. Handy, and the company was called Handy & Wetherell.

Witherell Scythe Co.   Oakland   1885-1926
Tools Made: Axes and Scythes
Remarks: Their mark **WITHERELL SCYTHE CO. | OAKLAND, MAINE** has been reported on a paper ax label. Another company, Witherell Axe Co., supposedly was in operation during the same years as this company in Oakland. Perhaps these two companies were the same and the name was changed at some point. Klenman (1997) indicates that James H. Witherell opened the company in 1886, and his sons, Carl and Louis, continued the business when he died in 1916. They sold it to Emerson & Stevens in the mid-1920s. E. & S. continued to use the label for awhile.

Withington & Hardy   Lewiston   -1869-
Tools Made: Brushes

Woltz, William G.**   Waldoboro   1856
Tools Made: Shipsmith
Remarks: He is listed in the 1856 *Maine Business Directory*.

Wood, Joseph**   Richmond   1735-1741
Tools Made: Gunsmith
Remarks: Demeritt (1973, 166).

Wood, Phineas**   Liberty   -1855-
Tools Made: Edge Tools
Remarks: He is listed in the 1855 *Maine Business Directory*. No other information is available about the tools he made or the location of his forge.

Woodbridge, R. R.**   Rockland   -1855-
Tools Made: Pump and Block Makers
Remarks: He is listed in the 1855 *Maine Registry and Business Directory*.

Woodbury*   Gray
Tools Made: Axes

Woodbury, A. K.*   Ellsworth   1891-1900
Tools Made: Axes

Woodcock, Nathaniel**   Thomaston   -1869-
Tools Made: Shipsmith
Remarks: He is listed in the 1869 *Maine Business Directory*.

Woodward, A. R.   Ellsworth   -1862-
Tools Made: Edge Tools

Woodward, E. W.**   Norway   1884
Tools Made: Gunsmith
Remarks: Demeritt (1973, 166).

Woodward, John**   Columbia   -1855-1856-
Tools Made: Edge Tools
 Remarks: He is listed in the 1855 and 1856 *Maine Business Directories* as an edge toolmaker.

Woodward, Nathan**   Brunswick   -1809-
Tools Made: Blacksmith

Wottrick, Rudolph**   Caribou   1886
Tools Made: Gunsmith
Remarks: Demeritt (1973, 166).

Wright, Joseph**   Jefferson   -1856-
Tools Made: Edge Tools
Remarks: He is listed in the 1856 *Maine Business Directory* as an edge toolmaker.

Wright, Js.**   Somerset County   1819
Tools Made: Blacksmith
Remarks: The Somerset County Probate inventories list him as owning a shot gun $15, a gun & bayonet $7, 1 share in the library, 2 Bibles, and Morses Geography.

Wrisley, Loren H.**   Norway   1835-1876
Tools Made: Gunsmith
Remarks: Demeritt (1973, 59-60, 166).

Young, Andrew**   Liberty, Montville   -1860-1870-
Tools Made: Blacksmith
Remarks: He is listed in Montville in the 1860 census and then in Liberty in the 1870 census.

Young, Charles L.**   Newport   -1878-1879-
Tools Made: Axes
Remarks: He is listed as C. L. Young in the 1879 *Maine Business Directory*. Tom Lamond notes that Charles L. Young of Newport has Pat. No. 199,342 for an improvement in axes, January 15, 1878.

Young, David**   Portland   -1869-
Tools Made: Shipsmith
Remarks: He is listed in the 1869 *Maine Business Directory* at 271 Commercial St.

Young, Dunbar H.**   Montville   -1870-
Tools Made: Blacksmith
Remarks: He is listed in the 1870 census.

Young, F. M.**   Camden   1878
Tools Made: Gunsmith
Remarks: Demeritt (1973, 166).

Young, George W.**   Montville   -1880-
Tools Made: Blacksmith
Remarks: He is listed in the 1880 census.

Young, Joshua**   Woodstock   -1855-1856-
Tools Made: Edge Tools
Remarks: He is listed in the 1855 and 1856 *Maine Business Directories*.

Young, Shepard**   Portland   -1869-
Tools Made: Shipsmith
Remarks: He is listed in the 1869 *Maine Business Directory* at 1 Cotton St.

Registry Information Sources

The principal references and information sources used to compile the registry listing are shown below in bold.

Adams, George. 1855. *The Maine Register for the Year 1855: Embracing State and County Officers, and an Abstract of the Laws and Resolves: Together with a complete Business Directory of the State and a Variety of Useful Information.* Boston.

American Genealogical Research Institute. 1972. *Walker family history.* Washington, DC: American Genealogical Research Institute.

Astragal Press. 1989. *The Stanley catalog collection: 1855 - 1898: Four decades of rules, levels, try-squares, planes, and other Stanley tools and hardware.* Mendham, NJ: Astragal Press.

Bacheller, Milton H., Jr. 2000. *American marking gages, patented and manufactured.* 185 South St., Plainville, MA 02762: Self-published.

Baker, William A. 1973. *A maritime history of Bath, Maine and the Kennebec region.* 2 vols. Bath, ME: Maritime Research Society of Bath.

Biographical Review. 1897. *Biographical Review, containing life sketches of leading citizens of Sagadahoc, Lincoln, Knox and Waldo Counties.* vol. 20. Boston, MA: Biographical Review.

Brack, H.G. 2008a. *Art of the edge tool: The ferrous metallurgy of New England shipsmiths and toolmakers 1607 – 1882.* Vol. 7 of *Hand tools in history.* Hulls Cove, ME: Pennywheel Press.

Brack, H.G. 2008b. *The classic period of American toolmaking 1827-1930.* Vol. 8 of *Hand tools in history.* Hulls Cove, ME: Pennywheel Press.

Branin, M. Lelyn. 1978. *The early potters and potteries of Maine.* Maine Heritage Series No. 3. Augusta, ME: Maine State Museum.

Brundage, Larry. n.d. Down East (Maine) planemakers. *Plane Talk.* 7. 1-20.

Brundage, Larry. 1981. Oakland, Maine axes and other edge tools. *The Chronicle.* 34.

Brundage, Larry. 1984. The planes of Maine. *The Chronicle.* 37. 21-4.

Brundage, Larry. 1989. An early and elegant broad axe. *The Chronicle*. 42. 97-8.

Butterworth, Dale, and Clarence Blanchard. 2005. Central Maine log rule makers and their rules. *The Fine Tool Journal*. 11-4.

Butterworth, Dale, and Bennett Blumenberg. 1991a. David Fuller, rural planemaker of West Gardiner, Maine part I. *The Chronicle*. 44. 67.

Butterworth, Dale, and Bennett Blumenberg. 1991b. David Fuller, rural planemaker of West Gardiner, Maine part II. *The Chronicle*. 44. 112-3.

Butterworth, Dale, and Bennett Blumenberg. 1992. Abel Sampson: Maine's privateer turned planemaker. *Maine Planemakers*. No. 3. Self-published.

Butterworth, Dale, and Bennett Blumenberg. 1993. E. S. Lane: A Maine rule maker and scaler. *The Chronicle*. 46. 15-6.

Clerk's Office of the District Court of the District of Massachusetts. 1867. *The Maine business directory for the year commencing 1867, a complete index to the mercantile, manufacturing and professional interests of the state, together with much valuable miscellaneous information.* Boston: Briggs & Co., Publishers.

Clerk's Office of the District Court of the District of Massachusetts. 1869. *The Maine business directory for the year commencing January 1, 1869, a complete index to the mercantile, manufacturing, and professional interests of the state, together with much valuable miscellaneous information.* Boston: Briggs & Co., Publishers.

Cohen, Marcie. 1988. The journals of Joshua Whitman, Turner, Maine, 1809-1846. In *The Farm*, ed. Peter Barnes, Boston: Boston University Press.

Cope, Kenneth L. 1994. *Makers of American machinist's tools: A historical directory of makers and their tools.* Mendham, NJ: Astragal Press.

Cope, Kenneth L. 1999. *American wrench makers 1830 - 1915.* Mendham, NJ: Astragal Press.

Demeritt, Dwight B., Jr. 1973. *Maine made guns and their makers.* Hallowell, ME: Paul S. Plummer, Jr. for the Maine State Museum.

Donahue, Tom. 1996. *The Kingdom in Montville, Maine: A technological diary 1789 - 1994.* Freedom, ME: self-published.

Eaton, Cyrus. 1851. *Annals of the town of Warren in Knox County, Maine with the early history of St. Georges, Broadbay and neighboring settlements on the Waldo Patent.* Hallowell, ME: Masters, Smith and Co..

Ellis, Alice. 1980. *History of Prospect 1759-1979.* Prospect, ME: Town of Prospect.

Fales, Cliff. 1992. The spiral screwdrivers of Isaac Allard, F. A. Howard and J. W. Jones. *The Gristmill.* 67. 10.

Getchell, Nancy L. 1956. *The manufacture of axes and scythes in Oakland, Maine.* Unpublished.

Godfrey, John E. 1882. *History of Penobscot county, Maine.* Cleveland, OH: Williams, Chase, and Co.

Goodman, W.L. 1964. *The history of woodworking tools.* New York: David McKay Company, Inc.

- The most important information source on the origins of woodworking tool manufacturing, their design, and the gradual invention of new tool forms.

Goodman, W.L. 1993. *British planemakers from 1700.* Mendham, NJ: Astragal Press.

Gordon, Robert. 1996. *American iron, 1607 - 1900.* Baltimore, MD: Johns Hopkins University Press.

Grindle, Roger L. 1971. *Quarry and kiln: The story of Maine's lime industry.* Rockland, ME: Courier-Gazette.

Grindle, Roger L. 1977. *Tombstones and paving blocks: The history of the Maine granite industry.* Rockland, ME: Courier-Gazette.

Hey, David. 1997. The development of the English toolmaking industry during the seventeenth and eighteenth centuries. In *Eighteenth-century woodworking tools: Papers presented at a tool symposium: May 19-22, 1994.* Gaynor, James M., ed. Vol. 3 of *Colonial Williamsburg Historic Trades.* Williamsburg, VA: The Colonial Williamsburg Foundation.

- Hey helps sketch in important background information about the origins of the skills of early American toolmakers. In contrast to unsigned American tools, the English production of cast steel tools can be carefully traced because all the producers, without exception, signed their tools or, at least, put their signatures on tools produced by their subcontractors.

Hoyt, Edmund S. 1875. *Maine state year-book and legislative manual for the year 1874-5*. Portland, ME: Hoyt and Fogg.

Hoyt, Edmund S. 1881. *Maine state year-book and legislative manual for the year 1880-1 from April 1, 1880 to April 1, 1881*. Portland, ME: Hoyt, Fogg and Donham.

Kallop, Edward L., Jr. 2000. *Johnson's kingdom: The story of a nineteenth-century industrial kingdom in the town of Wayne, Maine*. Wayne, ME: Wayne Historical Society.

Kallop, Edward L., Jr. 2003. *A history of the North Wayne Tool Co. manufacturers of axes, corn hooks, scythes and hay knives*. Wayne, ME: Wayne Historical Society.
- This text gives a comprehensive history of this company and the many names it used and people who were involved with it.
- "In the Maine Register the tool company's presence in North Wayne is first noted in 1881, when it is identified at Bodwell & Harvey, edge-tools. Not until the year following and thereafter is the published listing identified by company name. Whatever was the reason for the initial listing with personal rather than company name, it nevertheless leads to speculation on the complex role of William Harvey in these various transactions, and his emergence as an apparently equal partner with Joseph R. Bodwell" (Kallop 3002, 77).
- "At the opening of the new century the numerous firms manufacturing edge tools during much of the 19th century were reduced to three; the Dunn Edge Tool Company, Emerson & Stevens, and the American Axe and Tool Company. The last was to be out of business soon after the century began, leaving only two, but in 1907 they were joined by the King Axe Company. With an earlier existence as King & Messer, the company continued under its new name until 1922 when it was sold to others, then some twenty years later was resurrected and survived for a brief time as King Axe and Tool Company. With a far shorter lifetime is identified in 1906 still another newcomer to the list -- William Harvey & Sons" (Kallop 3002, 109).
- "In 1904, sharing a page with four others whose business addresses are in either Hallowell or Gardiner, is the North Wayne Tool Company. Identified as Manufacturers of Agricultural Edge Tools, the company's products are named under the heading Specialties: C. C. Brooks' Bread Knives, Corn Hooks, Hay Knives and Hoes. C. C. Brooks' little Giant Scythes. C. C. Brooks' Be Ve Be Scythes. H. S. Earle's Little Giant Grass Hooks. H. S. Earle's Corn Knives. Hand Made Axes of all Patterns. Lefavour's Favorite Weeders" (Kallop 3002, 110).

Kauffman, Henry J. 1972. *American axes: A survey of their development and their makers*. Brattleboro, VT: S. Greene Press.

Kebabian, John S. 1968. A visit to the Peavey factory site Oakland, Maine. *The Chronicle*. 21. 62-3.

Kennebec Journal. 1897. *History of Litchfield and an account of its centennial celebrations, 1895*. Augusta, ME: Kennebec Journal Print.

Kijowski, Gene W. 1990. *Directory of American tool makers: Colonial times to 1899: Working draft edition.* The Early American Industries Association.

Kingsbury, Henry D., and Simeon L. Deyo, eds. 1892. *Illustrated history of Kennebec County, Maine, 1625-1892.* New York: Blake.

Klenman, Allen. 1990. *Axe makers of North America.* Victoria, BC: Whistle Punk Books.

Klenman, Allen. 1997. The Witherell Axe Company of Oakland, Maine. *The Chronicle.* 50. 37.

Klenman, Allen. 1998. Josiah Fowler of New Brunswick. *The Chronicle.* 51. 25.

Kley, Ron. 1985. Researching early Maine craftsmen: John H. Hall and the gunsmith's trade. *Maine Historical Society Quarterly.* 24. 410-15.

Lamond, Thomas C. 1997. *The spokeshave book: Manufactured and patented spokeshaves & similar tools.* Lynbrook, NY: self-published.

Lamond, Tom. 2007a. Hubbard & Blake. *The Fine Tool Journal.* 57. 13.

Lamond, Tom. 2007b. Maine axe manufacturers. *The Fine Tool Journal.* 57. 12.

Lamond, Tom. 2008a. Emerson & Stevens Mfg. Co. *The Fine Tool Journal.* 57. 20-1.

Lamond, Tom. 2008b. Wm. Harvey & Sons. *The Fine Tool Journal.* 57. 18.

Lapham, William Berry. 1884. *History of Paris, Maine, from its settlement to 1880, with a history of the grants of 1736 & 1771, together with personal sketches, a copious genealogical register and an appendix.* Paris, ME.

Lowden, Linwood. 1984. *Ballstown - west - 1768-1809: An introduction to the history of the town of Whitefield, Maine.* n.p.

Lytle, Thomas G. 1984. *Harpoons and other whalecraft.* New Bedford, MA: The Old Dartmouth Historical Society Whaling Museum.

Mercer, Henry C. 1975. *Ancient carpenters' tools together with lumbermen's, joiners' and cabinet makers' tools in use in the eighteenth century.* Doylestown, PA: Bucks County Historical Society.

- The most important text on American 18th and 19th century tools. Well illustrated with an excellent description of the making of weld steel axes.

Merriam, Paul G., Thomas J. Molloy, and Theodore W. Sylvester, Jr. 1991. *Home front on Penobscot Bay: Rockland during the war years 1940 - 1945*. Rockland, ME: The Rockland Cooperative History Project.
Moore, Erwin W. 1940. *History and occupants of Main Street, Bingham around 1870*. Nancy Tancredi trans. from typed copy in the Bingham Library. Bingham, ME: Old Canada Road Historical Society.

Missouri Valley Wrench Club. 1987. *MVWC Newsletter*. 6.

Moore, Sam. 2000. Let's Talk Rusty Iron: Cant Hook or Peavey? *Rural Heritage*. Gainesboro, TN.

Moxon, Joseph. [1703] 1989. *Mechanick exercises or the doctrine of handiworks*. Morristown, NJ: The Astragal Press.

Muir, Diana. 2000. *Reflections in Bullough's Pond: Economy and ecosystem in New England*. Hanover, NH: University Press of New England.

Nelson, Robert E., ed. 1999. *Directory of American toolmakers: A listing of identified makers of tools who worked in Canada and the United States before 1900*. Early American Industries Association.

Office of the Librarian of Congress. 1855. *Maine business directory*. Boston: Briggs & Co., Publishers.

Office of the Librarian of Congress. 1879. *The Maine business directory for 1879. A complete index to the mercantile, manufacturing and professional interests of the state, together with much valuable miscellaneous information*. Boston: Briggs & Co., Publishers.

Office of the Librarian of Congress. 1881. *The Maine business directory, for 1881*. Boston: Briggs & Co., Publishers..

Office of the Librarian of Congress. 1882. *The Maine business directory, for 1882*. Boston: Briggs & Co., Publishers.

O'Leary, Wayne M. 1996. *Maine sea fisheries: The rise and fall of a native industry, 1830 – 1890*. Boston: Northeastern University Press.

Parks, Edward C. 1857. *The Maine register for 1857 with business directory for the year 1856.* Portland, ME.

Pearson, Ronald W. 1994. *The American patented brace 1829-1924: An illustrated directory of patents.* Mendham, NJ: Astragal Press.

Pierce, Cecil E. 1994. *Fifty years a planemaker and user.* Mendham, NJ: Astragal Press.

Pollak, Emil, and Martyl Pollak. 1994. *A guide to the makers of American wooden planes.* 3rd edition. Mendham, NJ: Astragal Press.

Pollak, Emil, and Martyl Pollak. 2001. *A guide to the makers of American wooden planes.* 4th edition. Revised and expanded by Thomas L. Elliott. Mendham, NJ: Astragal Press.

Rathbone, P. T. 1999. *The history of old time farm implement companies and the wrenches they issued.* Marsing, ID: P. T. Rathbone.

Rivard, Paul E. 1985. *Made in Maine: An historical overview.* Augusta, ME: Maine State Museum.

Rivard, Paul E. 2007. *Made in Maine: From home and workshop to mill and factory.* Charleston, SC: The History Press.

Roberts, Kenneth D. 1975. *Wooden planes in 19th century America.* Fitzwilliam, NH: Ken Roberts Publishing Co.

- The very first contemporary researcher on American planes, self-publisher of many important books about planes, and the first individual to widely publicize the forms and makers' signs on early American planes.

Robertson, Edwin B. 1977. *Maine Central steam locomotives: A roster of motive power from 1923 to the end of the steam era.* Westbrook, ME: self-published.

Robinson, Trevor. 1996. The ratchet screwdriver. *The Chronicle.* 49. 54-7.

- This article includes information on Maine toolmakers G. E. Gay, John Parsons, Zachary T. Furbish, F. L. Hamlen, and C. H. Mallett.

Rockland Bicentennial Commission. 1976. *Shore village story: An informal history of Rockland, Maine.* Rockland, ME: Courier-Gazette, Inc.

Rogers, Lore A., and Caleb W. Scribner. 1967. The Peavey cant-dog. *The Chronicle*. 20. 17-21.

Rosebrook, Donald, and Dennis Fisher. 2003. *Wooden plow planes: A celebration of the planemakers' art*. Mendham, NJ: Astragal Press.

Sawtell, William R. 1982. *Katahdin Iron Works: Boom to bust*. Milo, ME: Self-published.

Sawtell, William R. 1983. *Katahdin Iron Works revisited / compiled by William R. Sawtell*. Milo, ME: Self-published.

Sawtell, William R. 1988. *K. I. III*. Bangor, ME: Furbush-Roberts Printing Co.

Sawtell, William R. 1993. *Katahdin Iron Works and Gulf Hagas: Before and beyond*. Milo, ME: Self-published.

Sawtell, William R. n.d. Video: *History of Katahdin Iron Works and Gulf Hagas*.

Schulz, Alfred, and Lucille Schulz. 1989. *Antique and unusual wrenches*. Malcolm, NE: self-published.

Schwarz, Chris. 2004. Lie-Nielsen chisels: Better than vintage. *The Fine Tool Journal*. 54. 14-16.
- "Thomas Lie-Nielsen took the balance and feel of the [Stanley] 750, but he made the blade using tough cryogenically treated A2 steel instead of carbon steel. He also replaced the ugly red-painted oak (I think it's oak) handle with a finely turned and finished American hornbeam version" (Schwarz 2004, 14).

Sellens, Alvin. 1990. *Dictionary of American hand tools: A pictorial synopsis*. Augusta, KS: self-published.

- The first and best of the dictionaries on American hand tools.

Sibley, John Langdon. 1851. *A history of the town of Union, Maine, to the middle of the nineteenth century*. Boston: Benjamin B. Mussey and Co.

Smith, Roger K. 1981. *Patented transitional & metallic planes in America 1827 - 1927*. 2 vols. Lancaster, MA: North Village Publishing Co.

Springer, Moses. 1849. Largest scythe manufactory in the world. *Maine Farmer*.

Stahl, Jasper Jacob. 1956. *History of Old Broad Bay and Waldoboro: Volume one: The colonial and federal periods*. Portland, ME: The Bond Wheelwright Company.

The Union Weekly Times. 1895. *Union, past and present: An illustrated history of the town of Union, Maine from earliest times to date*. Union, ME: The Union Weekly Times.

Waters, Henry. 1982. *Kings' Mills: Whitefield, Maine 1772-1982*. Self-published.

Whelan, John M. 1993. *The wooden plane: Its history, form & function*. Mendham, NJ: Astragal Press.

Wheeler, George Augustus. 1875. *History of Castine, Penobscot, and Brooksville, Maine including the ancient settlement of Pentagoet*. Bangor, ME: Burr and Robinson.

Wheeler, George Augustus. 1878. *History of Brunswick, Topsham & Harpswell, including the ancient territory known as Pejepscot*. Vol. 2. Boston: A. Mudge & Sons, Printers.

Wheeler, George Augustus. 1896. *Castine past and present*. Boston: Rockwell and Churchill Press.

Wheeler, Robert. 1993. Nathaniel Potter: Could he be our earliest planemaker? A plane talk special feature. *The Chronicle*. 46. 3-4.

Whelan, John M. 1993. *The wooden plane: Its history, form & function*. Mendham, NJ: Astragal Press.

Wing, Anne, and Donald Wing. n.d. *The case for Francis Purdew or granfurdeus disputatus*. Self-published.

- A most important essay on the European sources of American plane designs.

Wing, Anne, and Donald Wing. 1984. *Chronology of 18th century planemakers in southeastern New England*. Marion, MA: The Mechanick's Workbench.

- An excellent survey of New England's first planemakers.

Wood, Richard G. 1935. *A history of lumbering in Maine, 1820-1861*. Orono, ME: University Press.

Yeaton, Donald G. 2000. *Axe makers of Maine*. Rochester, NH: self-published.

Journals

The Chronicle. Early American Industries Association (EAIA), Elton Hall, 167 Bakerville Rd., South Dartmouth, MA 02748-4198. www.eaiainfo.org.

- For over 50 years, this monthly publication has been an important source of information on toolmakers of every description.

Fine Tool Journal. Antique & Collectible Tools, Inc., 27 Fickett Rd., Pownal, ME 04069. www.finetoolj.com.

- Among the most important sources of information on Stanley tools, collectible wrenches, and obscure tool-related topics of all kinds.

Sign of the JOINTER. Southwest Tool Collectors Association, www.swtca.org.

Appendices

A: Maine Planemakers Working before 1900

The following list of Maine planemakers working before 1900 has been extracted from our registry. Please contact the Davistown Museum if you can provide us with information about planemakers working before 1900 that are not in this preliminary listing.

Barrows, L. M.   Vassalborough
Batchelder, I.   Sebago   circa 1790
Bead & Morrill   Bangor
Benjamin, Samuel   Winthrop   -1790-
Bennett's Mills   Norway   -1864-
Bliss, T.   Bucksport
Bradford, Joseph   Portland   circa 1837-1884
Burrowes Co., Edward T.   Portland   1878-1928
Carpenter, F. E.   Portland
Cary, W. H.   Houlton
Chase & Co., (George)   Portland   -1841-1846-
Chase, C.
Chick, George   Portland   ca.1850
Cluff, J. L.   Skowhegan
Coffin, G. W.   Freeport
Cram & Co., L.   Bangor   ca. 1840
Crie & Co., H. H.   Rockland   1860-1914
Crown Plane Co.   Bath/S. Portland
Darling & Schwartz   Bangor   1854-1866
Dennison, Charles   Freeport   -1869-1871
Dow, W. N.   Head Tide   ca. 1850
Emery, Waterhouse & Co.   Portland   1871-1885
Evans, George Franklin Norway   ca. 1862-1864-   (b.1842, d.1904)
Flyn, John   Warren   -1740-1760-
Foster, J.   Bath
Fuller, C. & D.
Fuller, David   West Gardiner   -1829-1856-   (b.1795, d.1871)
Goodhue, J. G.   Bangor   ca. 1810-1820
Gray, Arthur   Naples   -1867-
Haines & Smith   Portland
Harmon, A.   Scarborough -1810-1830-

Harmon, Benjamin   Biddeford, Scarborough, Buxton

Harmon, I. Jr.   Buxton?   (b. 1791)

Harmon, P.   Biddeford

Harris, T. W.   Portland   ca. 1850

Haynes & Smith   Portland   -1869-

Hersey, J. L. B.   Portland   -1849-

Hilton, J.   Kennebunk Landing   -1810-1830-

Holmes, I. P.   Berwick   ca. 1850

Jewett, J. C.   Waterville

King, J. K.

King, S.   ca. 1720-1820

Larrabee, J. C.   Brunswick   circa 1780 - 1820

Leighton, E. C.   Winthrop

Lie-Nielsen Toolworks Inc.   Warren   1981-present

Lovejoy, Hubbard   Wayne

Marden, I. W.   Palermo   -1869-

Mechanic Falls   ca. 1850

Merrick, George G.   Thomaston   ca. 1840-1860   (b.1814)

Metcalf, Joseph   Winthrop   (b. 1756 d. 1849)

Moriarty, Mathew   Bangor   ca. 1872-1906   (d.1911)

Morrill, Benjamin   Bangor   -1832-1851-   (b.1789, d.1862)

Nutting, Nathan   Otisfield   -1850-1867   (b.1804, d.1867)

O., T.   Old Town   -1840-

Ouellet   East Freeport   -1820-

Parris, A.   Portland   -1800-1820-

Phillips, Russell   Gardiner   ca. 1870

Pierce, Cecil

Plummer, Silas   Lisbon Falls   -1850-1870-   (b.1822)

Richmond, John W.   -1856-   (b. 1828 d. 1866)

Rider, Perry B.   Bangor   -1834-1848-   (b. 1808)

Rider, T. J.   Thomaston   -1861-

Robbins, Leon   Bath   -2001

Sampson, Abel   Portland   1820-1846   (b. 24 Aug. 1790, d. 1883)

Sewall, S.   York   -1745-1760-   (b. 1724, d. 1814)

Seward, W.   Bangor

Snowdeal, C. T.   Thomaston

Soule, Lewis S.   Waldoboro   -1849-1854-   (b.1813)

Sparrow, J.   Portland   ca. 1827-

Spear, C. A.   Warren

Starr, R. C.   Thomaston   (b. 1779 d. 1862)

Storer, Joshua P.   Brunswick   -1854-1873-

Taber, Allen   Augusta   -1870-1880-   (b.1800, d.1882)
Tarbox, F.   Calais   1871-1880
Titus, Daniel Butters   East Union   -1850-1910-   (b.1829)
Todd, B.   Albany   -1820-
W., J. H.   Waldoboro   ca. 1850
Walker, Abiel F.   Alna   b. 1808 - d. 1875
Wallace, G. E.   Jackson
Waterhouse, Emery & Co.   Portland
Waterman, Thomas Waldoboro   1800-1850   (b. ca. 1775)
Washburn, Oscar   Portland   ca. 1840-1845-
White, James   South Portland   2001-
Winslow, James H.   Thomaston   -1842-1865-   (b.1817)
Unknown maker   Mechanic Falls   ca. 1850

More details about each of the above planemakers can be found in the *Registry of Maine Toolmakers* listings, which preceded this appendix. Additional citations are always welcome.

B: The History of Shipbuilding in Maine: 1607 - 1915

1607 - 1649

The first ship built in Maine was the pinnace *Virginia*, built at the Popham Colony in 1607. The two basic forms of ships are already evident: full square rigged ocean-going sailing ships of European design and small two-masted fore- and aft-rigged proto-schooners, also of European design (shallops, ketches, later, dogbodies and pinkies). Maine's resource-based economy is a triad of forest products, fisheries, and furs.

Ship Types	Routes	Cargos	Shipbuilding Tools
Shallops and ocean-going ships of European design	Coasting, trans-Atlantic	Fish, lumber, fur, and colonists' supplies	Adz, broad ax, pit saw, auger - mostly imported English edge tools

1650 - 1720

The English Civil War (1649 - 1660) disrupts English shipping to the colonies and stimulates a period of intensive colonial shipbuilding. The importance of fur in Maine's economy is replaced by an emerging market economy, whereby fish and lumber are traded for other needed commodities. The basic designs of ocean-going ships remain unchanged until the end of the 18th century: square-rigged merchantmen and brigs, partially square-rigged with foresails.

Ship Types	Routes	Cargos	Shipbuilding Tools
Slow changes in shallop design, Evolution of two-masted Chebacco boat (Essex, MA)	West Indies trade; growing coasting trade; Gulf of Maine fishermen begin the pattern of fishing and trading in the winter and farming in the summer	Fish and lumber to W. Indies; Molasses and coffee back to ME; cord wood on coasters for the next 150 years	Weld steel edge tools continue to dominate basic toolkits - mixture of imported English- and American-made edge tools

1720 - 1775

After the beginning of King Philip's War (1686) and the French and Indian Wars (to 1759), very little shipbuilding occurs in Maine, except in Kittery and York (1693 - 1714). Chebacco boats evolve into the square-sterned dogbody. The Chesapeake Bay region and the Hudson River area are the most important shipbuilding areas. Changing designs in Maine fishing schooners derive from the development of the Jamaican, Bermuda, and Baltimore sloops and schooners in the 18th century, all utilized by Caribbean privateers. The late 18th century Gulf of Maine culmination of these designs is the Marblehead schooner.

Ship Types	Routes	Cargos	Shipbuilding Tools
Chebacco boats, Dogbodies, Sloops, Ketches, evolution of the Marblehead schooner	Growing trans-Atlantic trade; vigorous coasting trade; continued importance of the W. Indies trade	Fish, masts, spars, boxes, shingles, staves; some colonial iron to England	Basic toolkits unchanged; first appearance of cast steel edge tools after 1760; weld steel edge tools still predominate

1775 - 1790

The American Revolution interrupts the growth of colonial shipbuilding industries and the robust trans-Atlantic trade. The robust Chesapeake Bay eastern shore upriver country yards decline after 1790 - 1800.

Ship Types	Routes	Cargos	Shipbuilding Tools
Baltimore sloop; first Pinkies(?); Merchantman: ocean-going ships and brigs	Privateers and British warships harass American ships on all trading routes; China trade begins in 1785	NE - W. Indies - Europe trading triangle: forest products, dried fish, tea, molasses, salt, coffee	Up and down river and tidal sawmills a main source of timber and wood products; proliferation of imported English cast steel edge tools

1790 - 1807

The war between France and England (1793) stimulates the rapid expansion of American trans-Atlantic shipping. Establishment of US custom districts in 1790 allows historical documentation of American trading patterns and the growing New England cod fishery. Many ships for the China trade are built in Maine for Salem merchants. Wiscasset briefly flourishes as a major lumber port (1794 - 1796).

Ship Types	Routes	Cargos	Shipbuilding Tools
Two-masted schooners, Pinkies for inshore fishing, full rigged ships, 1795: 3-masted Chesapeake schooner	Neutral trade - high risk but high profits; dangers from English press gangs and French privateers; China trade	Yellow pine brought in from southern forests; timber to Havana for Spanish ships	Imported English cast steel edge tools; Samuel Slater's first power loom built in RI in 1790; US production of timber framing tools begins

1807 - 1816

Jefferson's embargo of December 22, 1807, halts all trans-Atlantic trade for fourteen months. After a brief revival, the War of 1812 inhibits the growth of shipbuilding; and the blockade of 1812 - 1814 halts most coastal shipping.

Ship Types	Routes	Cargos	Shipbuilding Tools
Slowly developing schooner designs;	High risk on all routes; China trade	Lime, lumber, and salt fish	First totally automated textile mill at Waltham,

double-ended pinkies used for inshore fisheries	continues; lime production at Thomaston stimulates the production of brick buildings in coastal cities for the next 75 years	exported out of New England; the golden age of the cooper begins	MA, signals the beginning of the factory system. New England toolmakers utilize imported English cast steel for edge tool production.

1815 - 1840

The end of the War of 1812 (through 1815) marks the beginning of the expansion of shipbuilding in Maine. Federal bounties on codfish catches stimulate a growing fishery in New England. Slow sailing schooners characterize a relatively quiet period of maritime history. By 1830, Maine has surpassed all states except New York in shipbuilding tonnage.

Ship Types	Routes	Cargos	Shipbuilding Tools
Slow growth in the size of schooners; era of full rigged ships	Rapid growth of coastal cod fishery, coasters and coasting packets, and West Indies trade; China and trans-Pacific trades continue	Fish, house frames, cotton goods, and hand tools; coastal passenger service; molasses, mahogany from the W. Indies	First steam vessel, 1820. Circular saw, introduced in 1825, stimulates the more efficient production of lumber and ships' timbers. Small tool factories and foundries proliferate all over New England.

1840 - 1848

The rapid growth of the cod fishery and shipbuilding in eastern Maine in the 1830s culminates in a burst of shipbuilding activities in all areas of coastal Maine. Maine surpasses New York in shipbuilding tonnage. The most active years of shipbuilding on the Kennebec River north of Bath are 1840 - 1857.

Ship Types	Routes	Cargos	Shipbuilding Tools
Appearance of three-masted schooners; golden age of the full rigged ship began	Continued growth in now well-established trading routes; cotton an increasingly significant cargo	Cotton, lumber, fish, coopers' shooks; more molasses, coffee, some slave trading	First steam powered sawmills appear; spread of railroads throughout coastal New England begins

1848 - 1860

Discovery of gold in California revolutionizes ship design; and a sudden need for speedy ships results in the era of the clipper ship. The building of the clipper ships *Red Jacket*,

Typhoon, and *Ocean Herald* signal Maine's contribution to the clipper ship era. The largest number and tonnage of wooden ships are produced in Maine between 1847 and 1857, coinciding with the emergence of a vigorous New England edge tool producing industry. This is the era of mackerel seiners and gillnetters, and dory handliners replace deck handliners in the cod fishery. Block-makers, sail-makers, ships' blacksmiths, rope walks, etc., all proliferate with the shipbuilding industry.

Ship Types	Routes	Cargos	Shipbuilding Tools
Era of the Clipper ship, Sharp schooners, Three-masted medium clippers. Two-masted schooners used primarily for off-shore fishing and the continuing coasting trade.	Competition to reach CA gold fields via San Francisco typifies the sudden desire for speedy schooners and ocean transports and temporarily de-emphasizes cargo capacity	Gold miners, sugar, soap, kerosene, tools to CA; beginning of huge grain trade from CA, also hides, copper	1850s mark the beginning of the factory system of making guns with interchangeable parts; emergence of major American toolmakers; cast steel slicks, adzes, and timber framing tools produced in New England

1860 - 1865

The sudden collapse of the market for and usefulness of clipper ships occurs in 1856. The rise of the factory system stimulates the need for large ocean-going bulk cargo ships. The advent of the Civil War interrupts both coasting, trans-Atlantic, and trans-Pacific shipping patterns. The withdrawal of bounties on codfish marks the end of the era of the New England cod fishery.

Ship Types	Routes	Cargos	Shipbuilding Tools
Increasing use of steam-powered vessels on open ocean; New England coasting trade increases after the Civil War	Privateering and Confederate raiders threaten all New England shipping; decline of Atlantic and Caribbean trading	Restricted and risky transport of fish, lumber, cotton, military supplies, and arms	Widespread use of American-made cast steel edge tools; invention of the direct process Bessemer steel production ushers in era of drop-forged tools

1865 - 1890

The twilight of Maine's shipbuilding era begins with the proliferation of railroads, ocean-going steamers, and the opening of the Suez Canal in 1869. The huge expense of building and operating coal-fired steamers provides a continuing opportunity for competition from Downeasters, which are inexpensive to build and operate and which represent a final golden era of sailing ship efficiency and design. The first four-masted schooner, *William White*, is built at Bath, ME, (1880), which achieved its peak production of wooden ships in 1882, especially coasting schooners. The demand for coopers' goods begins to decline with the proliferation of factory-made products. The first five-masted schooner is built in 1888; six- and seven-masted schooners come only after 1900.

Ship Types	Routes	Cargos	Shipbuilding Tools
Era of the Downeasters (square rigged sailing freighters); the *Henry Hyde* represents the high point of Downeaster design	Importance of New England coasting trade and off-shore fisheries continues; Maine dominates shipbuilding in the US with trans-oceanic bulk haulers	Era of bulk cargos: cotton, lumber, ice, fruit, factory goods; Pacific lumber products to east coast	The classic period of American machinist tools and patented steel hand planes; cast iron power-driven machinery begins supplanting hand tools in larger shipyards

1890 - 1915

The decline in Maine wooden shipbuilding industries is accompanied by increased production of steel steamers at Bath after 1895. Ships' masts are now imported from the Pacific northwest. The last full rigged ships, built at Phippsburg in 1893. The first steel-hulled sailing ship is built in Bath in 1894.

Ship Types	Routes	Cargos	Shipbuilding Tools
Five- and Six-masted schooners (ocean freighters - sailing barges) dominate shipbuilding activities in Maine	Development of the racing Gloucester fishing schooners and Maine's huge ocean-going schooners mark the end of the shipbuilding era in New England.	Granite and coal dominate bulk cargo shipments; ice trade ends; New England coasting trade dwindles after 1900	Steam-powered winches and pumps, diesel motors, and telephones radically change the operation of sailing ships; first auxiliary schooner built in 1900

Suggestions for this chronological sketch (additions, comments, criticisms) are always welcomed.

C: A Guide to the Metallurgy of the Edge Tools at the Davistown Museum

Art of the Edge Tool
An Exhibition Opened in June 2007

The following steel- and toolmaking strategies and techniques were used for the forging of the edge tools included in The Davistown Museum's Exhibition *Art of the Edge Tool*.

I. Steelmaking Strategies 1900 BC – 1930 AD

1. Natural Steel: 1900 BC – 1930 AD

Natural steel was made in direct process bloomeries, either deliberately or accidentally, in the form of occasional nodules of steel (+/- 0.5% carbon content (cc)) entrained in wrought iron loups. Bloomsmiths deliberately made natural steel for sword cutlers by altering the fuel to ore ratio in the smelting process, producing heterogeneous blooms of malleable iron (0.08 to 0.2% cc) and/or natural steel (0.2 to 0.5 cc and higher) or by carburizing bar or sheet iron submerged in a charcoal fire. Manganese-laced rock ores (e.g. from Styria in Austria or from the Weald in Sussex, England) facilitated natural steel production. As a slag constituent, manganese lowered the melting temperature of slag, facilitating the more uniform uptake of carbon in the smelted iron. The Chalybeans produced the first documented natural steel at the height of the Bronze Age in 1900 BC, using the self-fluxing iron sands from the south shores of the Black Sea. Occasional production of bloomery-derived natural steel edge tools continued in isolated rural areas of Europe and North America into the early 20th century.

2. German Steel: 1350 - 1900

German steel was produced by decarburizing blast-furnace-derived cast iron in a finery furnace, and, after 1835, in a puddling furnace. German steel tools are often molded, forged, or cast entirely of steel, as exemplified by trade and felling axes without an inserted (welded) steel bit. Such tools were a precursor of modern cast steel axes and rolled cast steel timber framing tools. German steel shared the world market for steel with English blister and crucible steel until the mid-19th century.

3. Blister Steel: 1650 - 1900

Blister steel was produced by carburizing wrought iron bar stock in a sandstone cementation furnace that protected the ore from contact with burning fuel. It was often refined by piling, hammering, and reforging it into higher quality shear or double shear steel or broken up and remelted in crucibles to make cast steel. Blister steel was often used for "steeling" (welding on a steel cutting edge or bit) on axes and other edge tools.

4. Shear Steel: 1700 - 1900

Shear steel was made from refined, reforged blister steel and used for "steeling" high quality edge tools, such as broad axes, adzes, and chisels, especially by American edge toolmakers who did not have access to, or did not want to purchase, expensive imported English cast steel. The use of shear steel was an alternative to imported English cast steel for making edge tools in America from the late 18th century to the mid-19th century.

5. Crucible Cast Steel: 1750 - 1930

Crucible cast steel was made from broken up pieces of blister steel bar stock, which were inserted into clay crucibles with small quantities of carboniferous materials (e.g. charcoal powder). After melting at high temperatures, crucible cast steel was produced in 5 to 25 kg batches and considered to be the best steel available for edge tool, knife, razor, and watch spring production. Due to lack of heat resistant clay crucibles, extensive production of high quality crucible cast steel didn't begin in the United States until after the Civil War.

6. Brescian Steel: 1350 - 1900?

Brescian Steel was a common Renaissance era strategy used in southern Europe to make, for example, steel for the condottiers of the Italian city states. Wrought or malleable iron bar stock was submerged and, thus, carburized in a bath of molten pig iron. Brescian steel cannot be visually differentiated from German steel or puddled steel, both of which were produced from decarburizing pig iron.

7. Bulk Processed Steel: 1870 f.

After the American Civil War, a number of new strategies were invented for producing large quantities of steel, especially low carbon steel, that was required by the rapid growth of the industrial age and its factory system of mass production. The first important innovation was Henry Bessemer's single step hot air blast process, followed by several variations of the Siemens-Martin open-hearth furnace and electric arc furnaces. For edge tool production, the electric arc furnace supplanted, and then replaced, crucible cast steel in the early decades of the 20th century. A few modern drop-forged edge tools are included in this exhibition as examples of modern bulk process steel producing strategies.

For more information on these later techniques, including the drop-forging of the all cast steel ax, see:
Davistown Museum *Hand Tools in History* **series Volume 11:** *Handbook for Ironmongers: A Glossary of Ferrous Metallurgy Terms: A Voyage through the Labyrinth of Steel- and Toolmaking Strategies and Techniques 2000 BC to 1950*
This publication is available for hands-on perusal by museum visitors.
Also, see the exhibition handout: **Edge Toolmaking Techniques**

D: Woodworking Tools of the 17ᵗʰ and 18ᵗʰ Centuries

The basic woodworking toolkits of early New England settlers remained almost unchanged from the early years of the colonial period until the beginning of the 19ᵗʰ century. However, how the tools were made, who made them, and where they were made did change. New England's first shipbuilders often had access to only a limited selection of woodworking tools. The most important ones are shown in **bold** in the table below. As shipbuilding became more sophisticated in the 19ᵗʰ century, with larger ships of more complex design being constructed in or near communities with larger populations and toolmaking capabilities, any of the following tools could have been in a woodworker's toolkit. After 1840, the advent of steam power, the rotary sawmill, and cast iron machinery gradually supplanted the function of many ship carpenters' and other woodworkers' hand tools.

Later tools (post 1750) are enclosed in parentheses ( )

Axes	Adzes	Other Edge Tools	Other Hand Tools
felling	**lipped**	**gouge**	**caulking iron**
broad	peen (peg poll)	**framing chisel**	brace
mast	block (poll)	forming chisel	bow drill
hewing	coopers	mortising chisel	pump drill
mortising	bowl	corner chisel	spud
lathing hatchet	**Measuring Tools**	turning tools	pickaroon
shingle hatchet	**dividers**	**slick**	log dog
coopers	square (framing, try, etc.)	pod auger (screw auger)	cant hook
twybil (Penn Dutch)	level	spoke shave	pike
(ice)	chalk, line and reel	drawknife	peavey
Saws	marking gauge	mast shave	grappling hook
pit	traveler	coopers' shave	ring dogs
frame	plumb bob	wedge	barking iron
whip	calipers	scorp	saw set
buck (bow)	**Hammers and Mallets**	froe	shaving horse
cross cut	wood mallet	gimlet	block and tackle

hand saw	beetle		trowel
tenon	**caulking mallet**	timber scribe	screw clamp
fret	claw hammer	screw tap	(breast drill)
Planes		block knife	(mortising drill)
block		chamfer knife	
smooth			
fore (or trying)		**Other Trades**	
moulding or molding	**Blacksmith**	**Cobbler**	**Farrier**
joining	**Courier**	**Flax-dresser**	**Fuller**
plough or plow	**Weaver**	**Carriage-maker**	**Lime burner**
mitre			
shoulder			
squaring			

E: Abiel F. Walker, Planemaker of Alna, Maine

In September of 2001, the Davistown Museum was fortunate in obtaining a small collection (+/- 20) of Abiel Walker's planes directly from the attic of the house in which he lived most of his life in Alna, Maine. At first glance it appeared to be a group of generic run of the mill planes, which if sent to a local auction in Maine, would sell for $5 to $20. Upon closer inspection, we determined that they have a historical significance because many of them were signed by the person who made them. Due to the location in which they were found and the family provenance, the identity of Walker as a planemaker and craftsman is clearly known.

Abiel Walker was a carpenter, boat builder, and planemaker who spent most of his life in Alna, Maine (1808 - 1875). A number of the planes he made and signed can be compared with other planes in his tool collection that he owned but did not make, and a comparison of the planes in his small collection illustrate the change from the era of owner-made hand planes of the 18th and early 19th centuries to the factory-made hand planes that became prevalent with the mass production of companies such as the Union factory at Pine Meadow near New Haven, CT. Three planes in Abiel Walker's toolkit appear to be late 18th century tools and pre-date the planes that Walker himself made. These unsigned planes, which appear to be made of mahogany, are all by the same hand but vary in size: 8 5/8" long bead, 9 ½" molding, and a 10 3/8" bead. All have matching wedges, wood, patina, beveling, etc. Prior to making his own planes, Walker may have obtained these from another carpenter or boat builder in the Alna area who made his own planes. All are signed with Walker's initials "AFW" (photo).

The next grouping in Walker's toolkit is those planes that he apparently made himself. As well as the double sash, panel raising, and skew panel planes already noted in the Maritime III collection, there are three additional planes. One is a 9 ¾" long molding plane constructed out of beech; the other two are a particularly intriguing matched pair of tongue and groove planes made out of oak, with brass lower plates instead of the characteristic iron plate of the factory-made tongue and groove planes. They are clearly handmade and in Walker's characteristic wedge style. Walker must have seen and used the factory-made prototypes of these pistol grip tongue and groove planes, which were widely available by 1850 (e. g. T. Tileston, 1802 - 1860, A. Cummings, 1848 - 1854, H. Chapin, Union Factory, 1828 - 1860). For whatever reason,

Walker made his own tongue and groove planes that appear to be equal in quality to the factory-made specimens of the time.

In addition to the planes that pre-date the period of Walker's planemaking activities, other factory-made planes were found in Walker's toolkit. Of particular interest is a signed J. T. Jones, Philadelphia, 9 ¾" complex molding plane, also marked with Walker's initials. Pollak (2001) lists J. T. Jones as working in Philadelphia between 1831 and 1846, which is also the likely period of Walker's most productive planemaking activities. Also in Walker's toolkit is another signed Union Factory, H. Chapin, molding plane with boxwood spline, 9 3/8" long. Interestingly enough, this plane is not signed with Walker's initials, raising the possibility that it was added to his toolkit late in his career or after his death, circa 1875. Walker's toolkit also contained two other Union Factory planes, i.e. a quarter round molding and a small sash plane, both signed with Walker's initials. There were also three other early molding planes without initials and a number of undistinguished later molding planes.

No information is currently available about who may have used Walker's toolkit after he passed away. His small tool collection is significant in illustrating the activities of a small noncommercial planemaker who built tools for his own use and possibly for the use of others in the local community and who also made practical use of planes that obviously predate his active years. The appearance of factory-made tools manufactured after 1840 in Walker's toolkit is symptomatic of the rapid changes in technology and in social and mercantile relationships that characterized the period before the Civil War. While thousands of Maine ship, boat, and house carpenters continued to make their own planes throughout the 19th century, the increasing intrusion of factory-made planes in Maine toolkits and tool collections signals the rise of the factory system and the mass produced planes of both the wooden plane manufacturers noted above and the more practical steel planes of Stanley and other companies that soon supplanted wood plane production.

F: Reference List of Important New England Edge Toolmakers

Name	Birth - Death Dates	Working Dates (?)	City	VIP
Faxon (first name unknown)		+/- 1795	Braintree, MA (to Jesse Underhill 1824)	
The Underhill clan was a large family of toolmakers working in Boston and NH from 1775 until +/- 1900.				!
Josiah Underhill	(1758-1822)		Chester, NH	
Jesse J. Underhill	(1784-1860)		Boston then Chester, NH	
Jay T. Underhill	(1802-1839)		Boston then Chester, NH	
Flagg T. Underhill	(1804-1850)		Chester, NH	
Samuel G. Underhill	(1809-1885)	1829 - 1852	Boston	
George W. Underhill	(1815-1882)	1839-1852	Nashua, NH	
Rufus K. Underhill	(1819-?)	1840-1869	Nashua, NH	
Hazen R. Underhill	(1821-1898)	1842-1898	Derry and Auburn, NH	
Leonard White L. & I. J. White	(1810-1893)	1837-1928	Buffalo, NY	
Bailey, Chany & Co.		1867-1869	Boston (Bought by Stanley Rule & Level in 1869)	
Leonard Bailey		1855-1884	Boston, MA and Hartford, CT (Bailey and Chany, Stanley Rule & Level)	
Seldon Bailey		1872-	Woonsocket, RI (Joined Stanley Rule & Level	

		1880	in 1880)	
August Stanley & Co. Stanley Rule & Level Co. Stanley Works		1854-1857 1857-1920 1920 f.	New Britain, CT August and Timothy Stanley were making hardware in New Britain as early as 1831.	
Thomas H. Witherby Witherby Tool Co. Winsted Edge Tool Co.	(?)	1849-1850 1868 1890 f.	Millbury then Winsted, CT Little is known of the prolific career of Thomas Witherby; sometime around 1890, Witherby Tool Co. became the Winsted Edge Tool Co.	!
Buck Bros Charles Buck		1853-1972 1873-1915	Worcester, MA then Millbury, MA Millbury, MA	!
Charles Hammond		1869-1908	Philadelphia	
James Swan James Swan Co.	(b. 1833)	1877-1951	Seymour, CT James was reported working in CT as early as 1856.	!
A. J. Wilkinson		1842-1993	Boston In the later years, they were hardware dealers.	
Collins and Co.	(1780-1835)	1826-1957	Canton, CT, which became Collinsville, CT	
Stephen Jennings		1836-1853		
Russell Jennings		1853-1944	Deep River, CT	
Marble Safety Axe Co.		1898-1911	Gladstone, MI (may have made hatchets as early as 1883.) Became the Marbles Arms Mfg. Co. in 1911.	
Bemis and Call		1844-1930	Springfield, MA	
Pardon Hayes Merrill	(1788-1879)	1820	Hinsdale, NH	

Pliny Merrill P. Merrill & Co.	(1800-1869)	1848-1858	
George S. Wilder Merrill & Wilder G. S. Wilder C. E. Jennings Jennings Edge Tool Works Jennings & Griffin	1828-1900	1873-1883 1858-1866 1873-1883 1883-1885 1885-1891 1885-1900+	Hinsdale, NH Merrill sells to Wilder in 1866 A New Haven, CT, hardware firm purchases G. S. Wilder George Wilder became the manager
Wilder & Thompson		1866-1868	Hinsdale, NH
Richard Henry Hopkins Wilder & Hopkins	1831-1877	1870-1873(?)	Hinsdale, NH
Oliver Sawyer	1794-1836	1808-1836	Bolton, MA

G: Reference List of Maritime Canadian Toolmakers

This listing is limited to edge toolmakers and planemakers. Most of these maritime peninsula Canadian toolmakers are from the *Directory of American Toolmakers* (Nelson 1999). Additional information on Nova Scotia and New Brunswick toolmakers not in this listing would be greatly appreciated.

Andrews, John D.   Milltown, NB   1889-1893
>*Tools made:* Axes and Edge Tools
>*Remarks:* Might be the same as John A. Andrews, son of John D. Andrews & Son.

Blenkhorn & Sons   Canning, NS   1866-1962
>*Tools made:* Axes and Edge Tools
>*Remarks:* Sons Sydney & Henniger joined him in his business.

Bradley Axe Factory   Nashwaak, NB   1862-1928
>*Tools made:* Axes, Handles, and Picks
>*Remarks:* Mark: a triangle with a dot in the center. Son George Todd Bradley succeeded him. Naswaak is now Durham Bridge.

Broad & Co., Elisha   Milltown, NB   1871-1883
>*Tools made:* Adzes, Axes, and Edge Tools
>*Remarks:* Elisha was not known to work alone, and in 1871 - 1883 he used the mark **E. BROAD | MILLTOWN**. In 1883, the company name was changed to E. Broad & Son.

Broad & Sons, Elisha   Milltown, NB   1883-1895
                                  St. Stephen, NB
>*Tools made:* Edge Tools
>*Remarks:* In 1883, Elisha Broad & Co. changed their company name to E. Broad & Son. In 1885, he added a second son and changed the name to Elisha Broad & Sons. One of the sons was named Harry W. In 1885, this company also moved from Milltown to nearby St. Stephen and used the old Douglas Axe Mfg. Co. factory located there. After Elisha died in 1895, they became the St. Stephen Edge Tool Co.

Broad, E. & J. W.   St. John, NB   1857-
>*Tools made:* Edge Tools
>*Remarks:* The Broad clan of edge toolmakers, including axes, was probably the most prolific of all maritime Canada edge toolmakers. Klenman (1990) provides an

extensive description of the Broad family ax business. Both their axes and edge tools still make occasional appearances in New England tool chests and collections. DATM (Nelson 1999) states that the E. is for Elisha Broad.

Broad, Elisha & Hewett   St. John, NB   1862-1866
    Tools made: Axes and Edge Tools
    Remarks: E. & J. W. Broad were succeeded by **E & H BROAD | CAST STEEL | ST. JOHN, N.B.** from 1862 - 1866. This was a partnership of Elisha and Hewlett Broad, who were brothers.

Broad, Hewett   St. John, NB   1881-1901
    Remarks: It is unknown what Hewlett did immediately after 1866, but, from 1881 - 1901, Hewlett used the mark: **H.BROAD | ST. JOHN.** John Harper has provided these photographs of a curved slick marked **H. BROAD | S^T JOHN N B | CAST STEEL**.

Burpee, I. & F. / I & E. R. St. John, NB   1862-1899
    Tools made: Edge Tools
    Remarks: Marks: **I. & F. Burpee**

Campbell & Fowler   St. John, NB   1863-1877
    Tools made: Edge Tools

Campbell Bros.   St. John, NB   1891-1920
    Tools made: Adzes, Axes, and Chisels

Campbell, William   St. John, NB   1881-1891
    Tools made: Axes and Edge Tools
    Remarks: Along with Josiah Fowler, William Campbell was one of maritime Canada's most important edge toolmakers. As Campbell & Fowler, William Campbell began working with Josiah Fowler at St. John, NB, as early as 1863.

Conolly   Middle River, NS 1877-
    Tools made: Axes

Cooper, William   Fredericton, NB   1865-1866
    Tools made: Edge Tools
    Remarks: A blacksmith who made edge tools.

Cox, Thomas   Kentville, NS   1882-1916

Tools made: Adzes, Axes, Coach-making Tools, Farm Tools, Knives, and Plane Irons

    Remarks: Marks: **PERKS/TC** (with cat's eye figure).

Drury, Edward   St. John, NB   1865-1866
    Tools made: Tools

Drury, Edward   St. John, NB   1817-1862
    Tools made: Wood Planes

Eaton, Benjamin   Cornwallis, NS   1865-1885
                            Sheffield Mills, NS
    Tools made: Adzes and Edge Tools

Eaton, George Wiswell   Berwick, NS   1872-1904
    Tools made: Axes and Edge Tools

Edwards, John C.   St. John, NB   1851-1884
    Tools made: Edge Tools

Elliott, J.   St. John, NB   1896-
    Tools made: Tools

Foster, G. P.   Darmouth, NS   1776-1811
                            Halifax, NS
    Tools made: Edge Tools and Other

Fowler Co. Ltd., Josiah   St. John, NB   -1860-1922
    Tools made: Adzes, Axes, Chisels, Drawknives, Hammers, and Picks
    Remarks: Josiah Fowler was one of the most important Gulf of Maine region edge toolmakers working in the 19th century. His high quality lipped adzes are still avidly sought by shipwrights and collectors. Klenman (1998) indicates he was working as a toolmaker as early as 1860, before serving in the Civil War as a bugler. After his return he was a partner in Campbell & Fowler, edge toolmakers who also made carriage springs and axles. He apparently worked under that name until 1877. The DATM (Nelson 1999) lists his dates as 1881-1920, but judging from the large number of lipped adzes and chisels that still turn up in New England tool collections and workshops Fowler was probably making edge tools much earlier than 1881 and may have been signing his tools with his own name while still associated with William Campbell, who was his probable partner in Campbell & Fowler. Signed lipped adzes, the tool for which Fowler was most famous, are uncommon. In contrast, lipped adzes with his characteristic touchmark, a dot

that sometimes looks almost like a triangle, are much more common. Fowler also used a number of other marks on his tools including "Fowler St. John", "Josiah Fowler St. John New Brunswick", and "Josiah Fowler Co." Fowler apparently was still active as late as 1922; his exact birth and death dates are not available. A more extensive information file is on the Davistown Museum website at http://www.davistownmuseum.org/bioFowler.html.

Harris & Allan   St. John, NB   1860-1888
    Tools made: Edge Tools

Harris, James Stanley   St. John, NB   1823-1860
    Tools made: Edge Tools

Hennigar, Henry   Milltown, NB   1881-
    Tools made: Saws

Hunt, Warren   Douglas, MA   1863-1892
                St. Stephen, NB
    Tools made: Axes

Johns & Co., H.   Lepreau, NB   no dates
    Tools made: Axes

Lenoir, Thomas   Arichat, NS   1880-
    Tools made: Axes

Maritime Edge Tool Co. St. Stephen, NB 1990-1911
    Tools made: Axes and Edge Tools

O'Connor, Thomas   St. John, NB   1881-
    Tools made: Cutlery
    Remarks: O'Connor was a gunsmith, whitesmith, and cutler.

Selig, William F.   Lunnenberg, NS 1870-1887
    Tools made: Edge Tools

Shields, Andrew   Halifax, NS   1818-1820
    Tools made: Axes

Spiller Bros.   St. John, NB   1867-1889
    Tools made: Axes, chisels, and Hammers

Remarks: The brothers succeeded their father, Samuel Spiller. A grandson of Samuel formed M. D. Spiller & Son Co. in Oakland, ME (1926), changing the name in 1937 to Spiller Axe 7 Tool Co.

Spiller, Hanford B.　　　St. John, NB　　　1862-
　　Tools made: Edge Tools

Stiller, C.**　　　　　St. John, NB　　ca. 1840-1850?
　　Tools Made: Edge Tools
　　Remarks: The Davistown Museum has a 4" wide slick (ID# 030505T1) marked **C. STILLER | ST. JOHN | CAST STEEL WARRANTED.** This slick is unusual; instead of being flat across, it has a central ridge with slightly slanted sides but no beveling. It came from a Brookline, MA, collector and is typical of the ship carpenter's edge tools that would have been used and traded throughout the Gulf of Maine region in the early to mid-19th century. Many Maine shipwrights traded or utilized edge tools by Josiah Fowler and other St. John makers.

Willis　　　　　　　St. John, NB　　　1865-1866
　　Tools made: Wood Planes

Wilson & Sons, Walter　St. John, NB　　　1877-1922
　　Tools made: Saws

H: Whalecraft Manufacturers of New Bedford

Name	Working Dates
James Barton	1852 - 1886
Frank E. Brown	1879 - 1923
Peleg Butts	1822 - 1849
(Taben & Butts)	1822 - 1873
Peleg Butts Jr. (son)	1845 - 1849
Butts & Smith	
William Carsley	1826 - 1843
William Carsley in Fairhaven	1846 - 1856
Luther Cole	1839 - 1859
Luther Cole in Fairhaven	1859 - 1892
Edward R. Cole (son)	1880 - 1924
Patrick Cunngler	1876 - 1885
Henry N. Dean	1839 - 1845
as Snell & Dean	1845 - 1856
as Dean & Sawyer	1856 - 1865
	1867 - 1897
Joseph G. Dean	1836 - 1847
as Dean & Driggs	1847 - 1876
James Driggs	1839 - 1902
James Durfee	1800 - 1845
Thomas Durfee	1810 - 1825 (?) 1831-
James Durfee Jr.	1820 - 1852-
Zoheth S. Durfee	1856 - 1861
Darius Gray	1838 - 1856
Daniel Kelleher	1869 - 1908
Josiah Macy & Sons (Josiah Jr., Edwin B. & Frederick)	1843 - 1904
James Maguire	1852 - 1894
Joseph B. Morse	-1852 - 1858

Howard Nichols	-1836 - 1852
Ambrose J. Peters	1879 - 1918
Charles E. Peters	1902 - 1925
Ebenezer Pierce	-1865 - 1902
Nathaniel S. Purrington	1836 - 1868
George P. Read	-1859 - 1880-
Lewis Russell	1835 - 1865
John A. Sawyer	1836 - 1896
Jonathan Smith	1836 - 1856
Alden G. Snell	-1836 - 1856-
James M. Snow	1839 - 1885
Lewis Temple	-1828 - 1854
Lewis Temple Jr.	1854 - 1858

The information above is from: Lytle, Thomas G. 1984. *Harpoons and other whalecraft*. New Bedford, MA: The Old Dartmouth Historical Society Whaling Museum.

I: English Edge Tool and Plane Iron Makers

English edge tool and plane iron makers are included as an appendix because they make such frequent appearance in New England workshops and tool collections. Many Maine planemakers utilized plane irons manufactured in Sheffield or Birmingham. A brief listing of the most commonly encountered names and marks helps clarify the fact that a large majority of domestically made planes utilized imported English plane blades. American-made plane blades, such as those manufactured by the Buck Brothers were infrequently used on American-made planes before the Civil War. The listing of English edge toolmakers, most of whom made plane blades, is also useful in helping to identify signatures on edge tools, especially drawknives and carving tools, which were made in England rather than by the principal edge toolmakers of Maine, New England, and New York. If not listed in the "Reference List of Important New England Edge Toolmakers" appendix F, additional American edge toolmakers are noted in the listings of tool manufacturers in vol. 8 of the Davistown Museum's *Hand Tools in History* series, *The Classic Period of American Toolmaking 1827-1930* (Brack, 2008b).

Addis (20th century)
George Bishop (19th century)
Butcher (Sheffield 1824 - 1900)
James Cam (Sheffield +/-1781 - 1838)
Joseph Cam (1860 - 1900)
William Greaves (Sheffield dates?)
Issac Greaves (Sheffield +/-1833)
John Green (Sheffield 1774-1824)
R. Groves (Sheffield 1824-1889)
Thomas Hall (19th century)
William Hall (Sheffield 1833-1881)
J. Herring & Sons (Sheffield? active 1940s)
James Horwarth (Sheffield 1872-1939)
William Hunt (Birmingham 1905- 1952)
Thomas Ibbotson (Sheffield 1825-1909)
Phillip Law (Sheffield 1787-1833)
William Marples (early 20th century - planes)
Marsh Brothers (Sheffield 1758-1960)
Moulson Brothers (Sheffield 1824-1912)
Samuel Newbould (Sheffield 1787-1881)
Edward Preston & Sons (19th century - planes)
Issac Sorby (Sheffield +/-1810)
John Sorby (Sheffield 1824-1881)

{Peter S. Stubs (Lancashire)} [Files, clamps]
Charles Taylor (Sheffield +/-1885)
David Ward (Sheffield 1824-1859)

1892077

Made in the USA